REAL QUANTA

REAL QUANTA

Simplifying Quantum Physics for Einstein and Bohr

MARTIJN VAN CALMTHOUT

DUNDURN
TORONTO

Translation by Tessera Translations

First published in 2015 by Uitgeverij Lias BV in Dutch. Published in English in 2017 by Dundurn Press Limited.

Cover image: istock.com/exdez
Chapter-opening illustrations: Wietse Bakker
"The Spirit of Delft" illustrations (pp134–39): Erik Kriek
Printer: Webcom

Library and Archives Canada Cataloguing in Publication

Calmthout, Martijn van
[Echt quantum. English]
Real quanta : simplifying quantum physics for Einstein and Bohr / Martijn van Calmthout.

Translation of: Echt quantum.
Translated by Tessera Translations.
Includes bibliographical references.
Issued in print and electronic formats.
ISBN 978-1-4597-4049-5 (softcover).--ISBN 978-1-4597-4050-1 (PDF).--
ISBN 978-1-4597-4051-8 (EPUB)

1. Quantum theory--Popular works. I. Title. II. Title: Echt quantum. English.

QC174.123.C3413 2018 530.12 C2017-906441-X
C2017-906442-8

1 2 3 4 5 21 20 19 18 17

Conseil des Arts du Canada
Canada Council for the Arts

We acknowledge the support of the **Canada Council for the Arts**, which last year invested $153 million to bring the arts to Canadians throughout the country, and the **Ontario Arts Council** for our publishing program. We also acknowledge the financial support of the **Government of Ontario**, through the **Ontario Book Publishing Tax Credit** and the **Ontario Media Development Corporation**, and the **Government of Canada**.

Nous remercions le **Conseil des arts du Canada** de son soutien. L'an dernier, le Conseil a investi 153 millions de dollars pour mettre de l'art dans la vie des Canadiennes et des Canadiens de tout le pays.

Printed and bound in Canada.

VISIT US AT

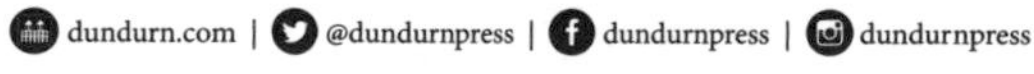

Dundurn
3 Church Street, Suite 500
Toronto, Ontario, Canada
M5E 1M2

CONTENTS

"Our brains are not built for quantum mechanics.
Trying to imagine it is pointless.
What we can imagine is classical mechanics.
Which is basically a weird kind of quantum mechanics."

Leonard Susskind

"If you think you understand quantum mechanics,
you don't understand quantum mechanics."

Richard Feynman

FOREWORD

QUANTUM 2.0, THE CONSTRUCTION AGE

Laws of nature are often born out of necessity, when you are unable to describe new, better observations using the known theories. Quantum theory was born a hundred years ago because the great physicists of the time, Bohr and Einstein, could not describe why atoms stick together, for instance, or have a particular color. Something new was needed, and that was quantum theory.

In the century of Quantum 1.0, the focus was on describing nature. Quantum theory proved to be so spectacularly good at it that nobody doubts the validity of the theory nowadays. Which is actually surprising, given that quantum theory is most definitely somewhat bizarre. The famous American physicist Richard Feynman put it succinctly: "The theory of quantum electrodynamics describes Nature as absurd from the point of view of common sense. And it agrees fully with experiment. So I hope you accept Nature as She is – absurd."

Just try to describe quantum theory in everyday language. Explain how a particle can be in two places at once or in two states at the same time (this is called 'superposition' in quantum-speak). Objects do not exist unless you look at them and consequences occur without a cause. And then there is the phenomenon of 'entanglement' – something happening elsewhere, instantaneously and regardless of the distance, when you do something here. Concepts such as superposition and entanglement are too weird for words.

Quantum theory was revolutionary not only for physics but also for philosophy. Everything great philosophers from Plato to Kant had said about 'reality' needed to be reconsidered. Our conceptual framework had to be extended if we were to be able to understand our universe again, even a little. That philosophical aspect is also part of Quantum 1.0, in which we try to understand nature – in all its absurdity – as it has turned out to be.

Now we are on the threshold of a new age, Quantum 2.0 if you like, which poses a thrilling question: "Now that we understand it better, can we use it to make things?" Can we build devices in which the components exhibit quantum behavior? Just imagine a car that is parked in two different places at once. Turn a steering wheel in Delft and a car in New York goes left. Superposition and entanglement create possibilities that could be incredibly efficient and useful. However, we will not be entangling cars any time soon. Not merely because they are far too big, but because the right components have not been designed yet.

To build quantum devices, we need a radically different way of thinking. Scientists started designing quantum building blocks twenty years ago. For a decade, various working building blocks have been made in a dozen laboratories scattered throughout the Western world. These building blocks consisted of single atoms at first, but we are now making bigger and bigger structures. Both the components and the 'device' as a whole demonstrate effects such as superposition and entanglement. Although this is still on a very small scale, we are getting more and more insights into devices that can do something revolutionary and new.

It is hard to predict how this potential for revolutionary technology will change our lives, but we are thinking hard about it, of course. The applications are for instance aiming to make detectors that can 'photograph' the magnetism of several atoms, which is very important for identifying the complex molecules in our body and could therefore be significant for medicine.

A great deal of attention is also being paid to secure communications. At the moment, the data center sees all the e-mails and pin codes you send. The vulnerability of this setup regularly makes the news. Quantum theory makes it possible to 'teleport' your pin code instead of sending it through a cable, and teleportation makes sneaky eavesdropping impossible. This is no longer science fiction: we already have teleportation over a distance of a hundred kilometers. I foresee that we will have a worldwide teleportation network within a few decades, significantly upgrading today's Internet.

Computing is perhaps the best example. We do that currently with bits of zeroes and ones, but quantum computing does calculations with superpositions of zeroes and ones, so information is coded as both zero and one, or both yes and no, at the same time. Computer scientists have already proved on paper that quantum calculations can be many times faster for a number of mathematical problems. Some of the biggest mathematical problems, ones that are impossible to solve at the moment, would be a piece of cake for a quantum computer.

A revolutionary computer like that would of course be interesting for all the large it companies. Over the past year, Google, ibm, and Microsoft have announced large investments in the development of such a quantum computer. Actually building one will be a massive milestone in technological development. Quanta will become part of everyone's daily life, there's no doubt about it.

We will have to begin explaining Quantum 2.0 in everyday language, as far as we can. This book by Martijn van Calmthout is a wonderful start. It makes a link between the history of Quantum 1.0 and the future of Quantum 2.0. I would have loved to contribute, but writing is an art form. And Martijn has mastered it.

Leo Kouwenhoven
Professor at Delft University of Technology
and scientific director of QuTech

THE MISUNDERSTANDING

Physics is the simplest subject there is. You don't really have to know anything. A few basic laws and a handful of math is enough. You can use that to calculate basically everything else, from the Big Bang to the light on your bike. It's not easy, but it's clear enough.

That simplicity is actually the main reason I started studying physics. I was, as I like to tell people at parties when they look at me in awe because I studied physics, just too stupid for other subjects such as history or French. Chemistry was a borderline case, to be honest, because of all those elements. Biology was obviously way too complicated, with its Krebs cycle and its intestinal peristalsis. Just give me Newton. Or Einstein.

But what about the theory of relativity? That's usually the next question. Do you really understand why a moving clock slows down and why you can never catch up with light, no matter what?

That's an easy one: There is no why! It's more the other way around. Starting by assuming you can never catch up with light makes everything else in the universe fall into place beautifully. The clock thing is just a side-effect of physics' favorite panacea, the constant speed of light.

Okay. But what about quantum mechanics? That's usually the next question. Those particles and waves, particles that can be somewhere and elsewhere at the same time, that can be infinitely far apart and still know what is happening to each other. That can't possibly be real?

Well, I tell them, you don't have to make sense of something to understand it. That's not the same thing. Making sense of something calls on your ability to imagine how something works. Understanding it is knowing what laws of physics apply in this case and how to work with those laws. It doesn't matter that an object can be both a particle and a wave at the same time. What matters is that you know that this explains how the world works.

I almost want to add that they don't have to worry about all that vague and confusing quantum stuff, because it is after all so far removed from the world as they know it. Particles do vague things, or have strange deeper interrelationships with each other. In everyday reality you don't actually have to bother about it.

But that, as I found out, is not quite true. The number of examples of quantum phenomena that do play a part in the tangible, everyday world is growing quickly. And so the idea of that wacky quantum world being more for professionals and enthusiasts than for ordinary people is getting less and less tenable. If you take a closer look, the quantum world is the one we really do live in. So the idea that we need not bother about it is a misconception, because we should. Because it's good to know how our world truly works. And because quantum reality is going to bring us some amazing technology.

This book is meant as a bridge between the incomprehensible quantum world of clinically cold laboratories and blackboards full of equations on the one hand, and everyday reality on the other. Those two turn out to be less far apart than it would seem, closer than physicists in particular conveniently like to claim. That last point especially is a downright shame. Real quanta present an excellent opportunity for physicists to bring their wonderful profession into the spotlight.

Amsterdam, 2016

OVERTURE

COFFEE WITH EINSTEIN AND BOHR

We've agreed to meet in the hotel bar, but I'm a bit early. Brussels is bustling busily by outside – it's about midday, often a hectic moment in Belgian social life. The Hotel Métropole, though, seems unaffected by the frenzied pace of life outside. I have entered the lobby, which looks like a hall of mirrors from the French Renaissance, a place from another time where lead glass, marble, cast iron and gold leaf are all entangled together to create characteristic Brussels Art Nouveau and Art Deco. A red-garbed bellhop hurries over, asking in barely a whisper what Sir requires.

The Métropole's brasserie, a little further into the hotel, is just as impressive as the lobby. Glittering cut-glass chandeliers illuminate a roomful of marble columns with semi-circular seats of red leather around them. On the other side of the round tables are comfortable armchairs in white leather. With their well-filled purses on a chain in the back pocket, the black-waistcoated and white-shirted waiters keep a discreet eye on the customers, gliding smoothly over to them when a beckoning hand is raised. People are mostly ordering coffee, or perhaps here and there their first beer.

This is where I have agreed to meet Albert Einstein and Niels Bohr, the two men who turned physics on its head a century ago. One of them, Einstein, had exposed a striking conspiracy between space and time with his theory of relativity: together they arrange things so that every observer sees light move at the same speed, no matter how fast the observers themselves are going.

That was in his *annus mirabilis*, 1905. Ten years later, in 1915, that conspiracy of space and time also transpired to be the key to a new theory of gravity that left nothing in the cosmos untouched.

Bohr, however, was the man for the very smallest scales, the atoms. In 1913, he suggested a radical new model for the structure of matter, proposing atoms as something like miniature solar systems in which electrons orbited a much heavier nucleus. This was able to explain a great deal about the way certain gases emit or absorb light. But the model came at a price: the electrons were only able to go round the nucleus in very specific orbits. Bohr didn't know why, but he knew it had to be true and he borrowed an idea that had been introduced to describe it in 1901 by his German colleague Max Planck. The energy of the electrons in the atom, he said, was quantized. Instead of being continuous, energy at the very smallest levels only occurs in packets of a fixed size. Planck himself had felt thoroughly uncomfortable with the concept, but as far as Bohr was concerned, Nature had clearly arranged it that way.

Their radical insights and ideas turned both these physicists into major figures at the start of the last century. And not just for their colleagues in the field. Einstein in particular was already a familiar name to the general public by 1927, a man whose compelling theories about the cosmos were regularly reported in the serious newspapers.

In the spring of 1927, here in Brussels, Einstein and Bohr both took part in the annual Solvay Conference, for which the formal theme that year was 'Light and Electrons'. In reality, the big names in physics used the conferences set up by Ernest Solvay to meet up once again in luxurious surroundings to discuss and exchange ideas. They were working, no doubt about it, but in a friendly and relaxed atmosphere.

That wasn't the case for Einstein and Bohr that year, however. Although they had been friends for years, they were attacking each other so vigorously that people around them were starting to find the debates uncomfortable. In the middle of their prolonged and protracted conversations, one of the pair would often get up and stalk off gesticulating, leaving the other behind, grumbling. The dispute could then start up again hours later, no matter what the time of day or what was officially on the schedule. Virtually every breakfast ended up being left untouched.

All these clashes between Einstein and Bohr were about the reality of quanta. A great deal had happened since Max Planck launched the idea in 1901, and indeed since Bohr had used it to explain the atom in 1913. Einstein had also been able to use it in 1905, for instance, to explain why light clearly behaved not only as a wave phenomenon but also as a kind of particle. And early in the 1920s, colleagues such as Max Born, Erwin Schrödinger, and above all Werner Heisenberg had set up an extensive theoretical network that seemed to describe the behavior of particles such as electrons and light down to the finest detail.

That description was what Einstein didn't like. The theory did admittedly yield excellent predictions for experiments, for example with electrons. But at the same time, he felt that something essential was missing: the insight. Heisenberg and the others could make excellent predictions of the relationships between variables. But Einstein saw that as little more than mathematical sleight of hand. The physical reality of what was going on deeper down in the system not only remained totally unclear, but also seemed to run counter to all intuition. To put it mildly. Particles were able to be in multiple places at the same time. They were both waves and particles. They could penetrate energy barriers.

Niels Bohr accepted the strange quantum phenomena as reality. Einstein kept asking him for deeper insights into the physics. Bohr stubbornly resisted that as a matter of principle, sometimes taking hours to think and then doggedly rejecting all the thought experiments Einstein could come up with. In the end, it seemed to be primarily a question of preference. God does not play dice, concluded Einstein. God's got nothing to do with it, reckoned Bohr.

That was in 1927. Now let's return to 2015. To Brussels.

It's taken me months of reading, talking and deliberation to work out what I'm going to tell the two of them once they're sitting at my cafe table on this historic spot. I asked them formally for an interview a couple of weeks ago – a dialog, actually, because I know that it's impossible to get a word in edgeways once Niels Bohr is talking unless you are as astute and as determined as Einstein.

But in fact, I'm not really going to interview the two of them at all. What I want to do is astound them. Astonish them with the aftermath of their own legendary discussions, the arguments that the pair of them had in 1927 at this very spot in Brussels.

The aftermath of their fight about what it actually meant, back when quantum theory was still in its cradle. A theory about energy and particles that was not yet fully fledged and that was at the very least an affront to decent scientific intuition, according to Einstein. Whereas Bohr, the younger of the two, kept telling Einstein that scientific intuition was perhaps simply inadequate for understanding the true nature of energy and particles. Accept the strangeness of quanta, he kept on repeating, and a new world will open up before your eyes.

Einstein and Bohr were guests in 1927 at the now famous annual Solvay Conference, organized by the Belgian industrialist and chemist Ernest Solvay. He was deeply interested in the latest insights in the sciences, and in 1911 he had invited the cream of international physics to a meeting at the Métropole Hotel in Brussels to discuss issues in modern physics. For the first few years, the Dutch theoretician Hendrik Lorentz from Leiden was the chairman. He was seen back then as the binding force in European physics.

The conferences in Brussels always included a group photo. The picture from 1927 shows this gathering of brilliant minds grouped on the terrace steps next to the Hotel Métropole. Einstein, gray-haired already, is sitting pontifically in the middle, with his typically unruly hair that made him a popular symbol of free-thinking genius even while he was still alive. Many young physicists still run their hands unthinkingly through their hair one extra time in the morning... On the far right of the second row, we see a younger-looking Niels Bohr, pale and rather introverted. He must have been tired after days of cutting-edge debate with one of the smartest physicists of his time.

SOLVAY CONFERENCE 1927
TOP A. Picard, E. Henriot, P. Ehrenfest, E. Hersen, T. de Donder, E. Schrödinger, E. Verschaffelt, W. Pauli, W. Heisenberg, R. H. Fowler, L. Brillouin

MIDDLE P. Debye, M. Knudsen, W. L. Bragg, H. A. Kramers, P. A. M. Dirac, A. H. Compton, L. de Broglie, M. Born, N. Bohr,
BOTTOM I. Langmuir, M. Planck, Mrs. M. Curie, H. A. Lorentz, A. Einstein, P. Langevin, C. E. Guye, C. T. R. Wilson, O. W. Richardson

Einstein is at the center and Bohr in the margins. That shows you the power relationships within physics at that time. But with hindsight, the Danish physicist Niels Bohr is really the central figure in the photograph. Bohr's ideas about the meaning of the new quantum theory gained more adherents in the years that followed than Einstein's eternal skepticism. That approach did indeed sharpen up his opponents' arguments, but ultimately he had to bow down in the face of reality: quantum reality is weirder than we could ever have thought.

Most physicists do seem able to live perfectly well with what has now become known as the Copenhagen Interpretation, after Bohr's Danish home base where virtually all talented young physicists stayed for a while to learn the trade. The Copenhagen Interpretation of quantum mechanics states in essence that it is pointless talking about anything other than measurable variables. Everything else, from deities rolling dice to hidden physical processes, is uninteresting as a matter of principle because by definition it cannot be measured.

I've gone through the history of quantum theory once again during the last few months, following a series of lectures by one of the contemporary grandmasters, and I've thought a lot about quantum mechanics. There is indeed something wrong with quantum theory, or at any rate with its hugely mysterious image. As soon as it starts going on about electrons that are in multiple places at the same time, or light particles that still sense each other even if they are a whole universe apart, then the old quote by the legendary physicist Richard Feynman springs to mind: "If you think you understand quantum mechanics, you don't understand quantum mechanics." And when people have understood that I myself am a physicist, you just know that they're going to ask if I understand it – all that stuff about particles and waves.

The mysterious image of quantum physics arises above all because we do not come across objects in our daily lives that are able to be in several places at once. Billiard balls, apples, spinning tops: everything has a clear position and a clear state. Everything moves according to laws of physics that we have a rough intuitive understanding of. Newton's laws, for instance, seem so understandable because they encapsulate mathematically what we can see in real life when things are moving and colliding.

As far as I'm concerned, the real mystery is why there is in fact a part of reality that doesn't behave according to the weird and wonderful laws of quantum theory. The real question is why an apple is not a quantum object, given that an electron is. Not the other way round.

We've been looking at the riddle of quantum mechanics from the wrong side for nearly a century now, and that's largely the fault of the physicists themselves. Big names such as Bohr and Einstein have presented the quantum world right from the start as the exceptional domain, far removed from everyday reality. Niels Bohr devoted most of his career to working out why we notice so little of the quantum world – nothing at all indeed – in our everyday reality. And Einstein kept on longing to find a classical underpinning at the bottom of the pit of quantum riddles.

Bohr's opinions about the classical and quantum worlds are not pointless; the subtle idiosyncrasies of the quantum domain easily get overwhelmed and averaged out by the sheer dimensions of large objects. What is left are then the age-old laws of physics that describe how objects move and even how gases contract and expand. But there are exceptions that permit quantum phenomena to play a demonstrable role in everyday existence.

That is the question that I now want to present to Einstein and Bohr. There are (I intend my opening gambit to be) good reasons for making the quantum world less of an exceptional domain than both of you so forcefully did for so long. To start with, so much has been discovered over recent decades that is new about the existence of quantum effects, even in the everyday world where you always thought it would be impossible. Amazingly enough, that's even the case to a significant extent in biology and biochemistry, fields involving warm aqueous mixtures of molecules that would seem to be the last place you would look for quantum effects, which generally give up the ghost at the slightest breath of wind or ripple of warmth. Nevertheless, the whole field of quantum biology is now based on it.

Even more importantly, your incessant squabbling about the exceptional nature of quantum reality, or in Einstein's case even the illogicality of it, has in fact robbed quantum mechanics of almost all goodwill among the general public, while it is nevertheless one of the deepest insights into the fundamental nature of our universe, our existence, and ourselves. Hammering on about the incomprehensibility of quantum reality means that you have each, in your own way, prevented the quantum way of thinking from becoming commonplace. Schoolchildren and students only tackled it if they really had to. Teachers stood trembling in front of the class. After all, anyone who said they understood it hadn't really got it... Thanks again, Richard Feynman.

That collective tremble is above all a missed opportunity, because quanta are playing an increasingly dominant role in our technology. Already, our smartphones, tablets, laptops, and iPads are full of technology that would have been inconceivable without quantum theory. And the end is nowhere in sight. Physicists and technicians throughout the world are working on techniques for using quantum phenomena directly for calculations on an unprecedented scale, communications that cannot be eavesdropped on, or encrypting confidential data. In the world of the future, the quantum aspects of nature will be more visible than ever. Mumbo-jumbo and magical mumblings will not be appropriate – at most there should be amazement at the coolly pristine beauty of quantum behavior. And the exceptional case of the laws of classical physics.

But above all, nevertheless, I have decided to listen to the responses of these two great thinkers to the marvelous insights and technology that their own theories have yielded in the meantime. From the unknown heights that solid-state physics has reached, from semiconductors in everything from silicon chips to solar cells, through to the magic of purely two-dimensional materials such as graphene. About the efforts to construct genuine quantum computers. About the cast-iron proof of what Einstein once referred to as the 'spooky action at a distance' between quantum particles. About teleportation of atoms. About light that switches magnets on and off and computer memory that no longer consists of minute magnetic domains but instead is made up of something much tinier still: individual electrons flipping their states, or spintronics as it is known.

And since we're on the subject, we will also talk about the amazing discoveries made in recent years about the crucial role that quantum magic has in the conversion of sunlight into sugars in green plants via photosynthesis. Or the way in which enzymes give structure to connective tissue, how we are able to smell individual molecules, the striking stability of our genetic material, or the fantastic quantum compass that some migratory birds seem to possess.

Until recently, biologists would acknowledge when pushed that chemistry is an essential part of their field. However, quantum biology has meanwhile become one of the key areas that young talented researchers dive into. It's possible that the quantum domain may even be the very crux of life. Getting to the bottom of that would also be the ultimate triumph of the wheels of quantum mechanics that Bohr and Einstein set in motion.

At the same time, there is the promise that fathoming out how Nature (which is one messy compromise after the other) nevertheless manages to utilize subtle quantum effects may also for instance introduce new ideas into technology. At the moment, quantum physicists and nano-engineers are ensconced in soundproof laboratories where they perform their experiments in high vacuum and at ultra-low temperatures. If they could only understand how a quantum compass consisting of just two electrons can exist in a European robin as it flies around the place, then they might also be able to apply their quantum effects in the tumultuous outside world as well. It should at least give someone like Bohr serious food for thought.

Movement, over by the swing doors. Here they come, the two great physicists. Einstein doesn't walk as easily as the rather younger Bohr, but they are both clearly still well with it, as if we were still somewhere around 1950. They're already gesticulating and talking. What else would you expect? German. English. Strong accents. The duty manager takes their coats and hats, and then points out the gentleman that they have come to visit, at the next table along.

I stand up and put my new smartphone down on the table. With its nanoscale circuitry and permanent connection to the entire Internet, the device is immediately one of the most tangible modern proofs that quanta are just as real as everything else around us. We'll be talking about that in detail.

But first I shake them cordially by the hand. "Gentlemen," I say. "Let's get started. Coffee? And shall we sit down?"

1
THE FIRST QUANTUM

IN WHICH MAX PLANCK ASSUMES THAT ENERGY IS IN SOME WAY GRANULAR, AND EINSTEIN AND BOHR DISCOVER THAT THIS IS NOT MERE MATHEMATICAL MISDIRECTION, BUT THE WAY LIGHT AND ATOMS WORK

It's an experiment that I once did as a student myself, but the result is so counterintuitive that I've looked it up once again just to be sure. I'm talking about the photoelectric effect, a phenomenon in which some metal surfaces acquire an electrical charge when light shines on them. The explanation of this effect is in fact what my table companion for this afternoon, Albert Einstein, received his Nobel Prize for in 1922.

It wasn't actually for his theory of relativity, which tied time and space together so innovatively but was also highly abstract, in his time at any rate. The theory had even been confirmed using astronomical observations, but otherwise seemed far removed from reality. There was no question that Einstein deserved the Nobel Prize, although the Nobel Prize Committee traditionally preferred to give the award to theories that had been proved by solid experiments in the laboratory. Einstein had one of those as well; one thing that was most definitely demonstrable, even back in 1922, was the photoelectric effect.

The experiment is straightforward. We'll take a thin sheet of metal connected to an ammeter, and a powerful light with a dimmer switch. Shine the light on the metal at full blast, and the meter gives a reading straight away. Charged particles are evidently being released by the metal; let's be very contemporary and immediately assume that these are electrons. Now let's find out what's happening. What is creating that current? Which parameters affect the results of the measurements and which don't?

We make measurements and record the observations in graphs.

What happens for instance if the light shines on the metal more brightly? The meter shows that more electrons are then emitted. It is also possible to show that those electrons are no more energetic than when the light intensity is low. What we do find, though, is that the energy of the electrons increases if the light shining on the metal is bluer. The graphs show that clearly and irrefutably.

Our scientific intuition is pretty bamboozled by this. Light, we were pretty certain, consists of waves. More intense light should have bigger waves. Bigger waves must be able to release more from the metal than small ones. But that's not what's happening here. What's going on? And even more strangely, why should waves with a shorter wavelength pump more energy than longer wavelengths into the particles that are emitted?

In 1905, Albert Einstein wrote no less than five scientific articles that made him an established name in physics in one fell swoop. Until that time, affectionately referred to as the *Wunderjahr*, Einstein was an absolute nobody. He had studied in Zürich, but without particularly convincing efforts or results, and he couldn't get a job anywhere. Job applications as far afield as to the Dutch physicist Kamerlingh Onnes in Leiden led to nothing. Einstein gave a few lessons and taught at a school for a while. But that was it.

Finally, he managed to wangle a position at the Patents Office at Berne in Switzerland through a friend's father. A third-level civil servant, and that would have to do for now. He left his pregnant companion Mileva Maric behind and went. Instead of feeling discouraged and uprooted, Einstein was enthusiastic about his relatively undemanding post, which gave him plenty of time to think about physics.

At the turn of the century, physics was in a deep crisis, although not everyone was aware of it. On the one hand, electricity and magnetism, light, gravity, pressure and temperature and numerous other physical quantities had been splendidly described in beautiful theories. But there were also open questions. What medium did the vibrations of light propagate in, now that experiments had shown that the putative ether did not exist? Another question was whether gases and solid substances really did consist of atoms, or was that no more than a theoretical crutch to help conceptualize things better? And of course there was also the photoelectric effect, which seemed to trample roughshod over all those nice ideas about light waves.

Despite being so young and inexperienced, or perhaps precisely because of that, Einstein was a pragmatist. His view was simple: if matter could best be described (by far) on paper as consisting of atoms, then atoms existed. In one of the articles from 1905, coincidentally also his thesis, he gave a straightforward calculation of how large such atoms would then have to be. He wasn't far off either, by modern standards.

But Einstein, busy during the daytime with patent applications from Swiss inventors and industrialists, kept on puzzling away at the physics of his time. The enigma of the photoelectric effect intrigued him and he then realized that there was an exceptionally simple explanation, that the light shining on the metal is essentially not a wave at all. If that light were to consist of packets of energy, like particles, you can immediately understand why more light releases more electrons, but not more energetic ones. The breakthrough came when he wrote down a formula that defined the relationship between the wavelength of light and the energy of the light particles. Light at the blue end of the spectrum has more energy per particle than red. So blue light imparts more energy to the electrons emitted from the metal than red does.

That may fly in the face of the idea of light as an electromagnetic wave, but it works. And the formula that Einstein used in 1905 for the relationship between the color of light and its energy was deliberately borrowed from professor of theoretical physics in Berlin, Max Planck. He had first published that expression five years earlier, in December 1900, in an article that with hindsight can be seen as the start of the quantum revolution. Like all other physicists, though, Planck himself hardly realized it at the time. When a young gun called Einstein used his formula in 1905 to prove that light consists of particles, he was in fact shocked. He emphasized that the energy formula was a calculation trick, not serious physics.

The story of Max Planck starts before the turn of the century and again involves a professional fraternal dispute. Planck was a serene figure, a professor in Berlin and a traditional thinker, specializing in the theories of Maxwell that describe electricity and magnetism. He was surrounded by the hubbub of the newcomers who talked about the atoms that matter and gases were thought to be made of. An Austrian by the name of Ludwig Boltzmann was particularly prominent, with controversial new theories about heat and entropy based on statistical descriptions of the motions of countless atoms. Planck wanted nothing to do with it. He thought that matter was a single electromagnetic whole. Heat was a question of electromagnetic vibrations, not quivering atoms. And atoms were only the theoreticians' fantasy anyway, not physical reality.

But the debate kept on going and Planck, usually averse to controversial actions, finally saw an opportunity around 1897 to deal with the atomists once and for all. He started doing calculations on the heat from a hot oven, a question that was getting a lot of attention at that time. In 1896, Wilhelm Wien had made measurements of the colors of light that were emitted from a closed oven at particular temperatures. The relationship he found was optimistically christened Wien's Law. One year later, though, the law turned out not to apply for infrared. The need was increasing for a good theory about the hot oven including that part of the spectrum.

What radiation would be present in the oven if it was left alone at a specific temperature? It sounds like an issue that has nothing to do with the existence of atoms and molecules, which made it ideal for Planck's attack on Boltzmann's atomists. If he was able to describe the hot oven without referring to atoms but including the deviations from Wien's Law, he would have struck a major blow. Those atoms would clearly have nothing to do with it.

On October 7, 1900, Planck started tackling the job in his villa in Berlin. As darkness fell outside, he stared on Wien's Law and made a few corrections to it, as theoreticians often do. It wasn't much more than guesswork, but the radiation formula that he found soon fitted in perfectly with all the infrared measurements from radiating hot bodies. All without involving any atoms vibrating and zooming about. It was a success.

In his derivation, Planck did introduce two constants without which physics has been inconceivable ever since. The first of them, written with a small letter h, has been known as Planck's constant for almost a century. The other, somewhat ironically, is Boltzmann's constant, written as a small letter k.

After the guesswork to find a suitable formula came the real work. What was the underlying physics for Planck's law of radiation? In order to get a better picture of the energy equilibrium, Planck saw the walls of the oven as abstract oscillators. They absorb waves in the oven, damp them down, and emit new waves. Equilibrium simply means that the processes are in balance. But calculating that wasn't so easy.

Finally Planck resorted to the statistical methods that he had hated so vehemently when Boltzmann used them. He later called it an act of desperation, but it worked. Planck was able to recreate his formula by assuming that the oscillators in the walls of the oven could only absorb or emit integer multiples of a small basic quantity of energy. And then he wrote down for the first time the formula that Einstein adopted literally five years later: radiation energy consisted of small packets, their size determined by the frequency of the light, i.e. its color.

In 1900, Planck himself was by no means ready for this. He was simply pleased that he was able to derive his law of radiation from first principles. But he did not know whether the imaginary oscillators in the oven walls actually existed. They were useful for the bookkeeping of the energy balance, he thought, but no more than that.

What had changed was his appreciation of Boltzmann. In his calculations, Planck gradually ended up using the statistical techniques that his Austrian colleague had been using for some time. What had started as an effort to pull the rug out from under the atomists' feet had done the exact opposite, and after 1900 Planck became an ardent advocate of the atomic theory of gases, heat, and matter. When measurements of hot bodies were done in 1913 in the highly energetic ultraviolet spectrum, they turned out once again to comply perfectly with Planck's law. The energy quantum had by then been on the rise for some years among theoreticians, although it was largely not understood. An uncomfortable idea, but one that regularly saved the day.

If there's one physicist whose day needed saving by the energy quantum, then it's the Danish physicist Niels Bohr. Around 1913, at the age of just twenty-eight, he became involved in the ongoing discussions about the structure of the atom. It was already pretty well accepted by then that matter consists of atoms. But what were these building blocks of matter made of, exactly? And what was their structure?

It seems completely obvious with hindsight, of course, the idea that an atom consists of electrons and protons and that those electrons are circling the nucleus in nice tidy orbits.

We're all too familiar with that kind of circular motion. The entire solar system consists of balls, the planets, describing ellipses around the sun at varying speeds. And we can see the same picture for instance in Jupiter and its dozens of moons. So anyone who realizes that an atom has a nucleus with electrons moving around it quickly spots that it is like a miniature solar system.

That's how it was around 1910 too. Insofar as physicists were interested in atoms at all – they preferred to think about radiation and waves – the discussions were about where the electrons were located in the atom. Ernest Rutherford had discovered with Geiger and Marsden in Manchester in 1908 that particles fired at extremely thin gold foil occasionally bounced back hard. They noted in amazement that it was as if rice paper had stopped a cannonball. But there in Manchester were the first seeds of the idea that an atom must therefore have a solid nucleus.

But if so, asked their colleagues elsewhere, where were the electrons? In some sort of negatively charged pudding with positively charged currants scattered throughout perhaps, as J. J. Thomson in Cambridge (UK) in particular proposed? Or in some kind of miniature solar system, as Rutherford and his colleagues intuitively imagined?

Again with the benefit of hindsight, it seems incomprehensible that the young Danish physicist Niels Bohr turned in Cambridge in the fall of 1911 to none other than Thomson, the man of the pudding model. This was the same Bohr who would publish three articles in July 1913 explaining with unprecedented precision exactly how an atom works and what it looks like: a miniature solar system, with the nucleus in the middle and electrons circling around it. Why didn't Bohr go to Rutherford in Manchester, if he already knew the answer? Why did he want to go to Thomson?

It looks strange now, but it wasn't. Niels Bohr didn't in fact go to England in 1911 to investigate atoms at all. He went there to tighten up his own work on electrons in metals, for which he had received a doctorate with distinction – *summa cum laude* – in May of that year in Copenhagen. There was one absolute authority in that field, the 1906 Nobel Prize winner John J. Thomson in Cambridge, the man he had to talk to.

But Cambridge was a disappointment. Bohr wrote desperate love letters to his young wife Margrethe saying that Cambridge was a mistake. In reality, Thomson turned out to have little interest in detailed discussions about electrons in metals and no interest whatsoever in the overly energetic and talkative Bohr, whose quite brilliant thesis had demolished Thomson's well-known theory about electrons. That had hurt and there was bad blood between them.

In his thesis, Bohr had proposed an innovative theory about metals in which he had used the then controversial ideas of the German physicist Max Planck, who had assumed that energy exists in small packages instead of being available in any desired quantity. That quantum theory turned out to be able to explain and predict a number of properties of metals that had remained enigmatic under the old theories of J. J. Thomson and others. As Einstein had done earlier, Bohr took Planck's statistical math trick at face value: energy really does come in packets.

Great idea or not, Thompson didn't want to talk about it and Bohr decided to make the best of a bad job. After a couple of months in Cambridge, he moved to Manchester where Ernest Rutherford and other big names in British physics were working. That meant that Bohr dropped his plans for improving his theory of electrons from the thesis. Instead, he became enthusiastically involved in Rutherford's research and investigations of the structure of the atom, in particular the location of the electrons within it.

His role was initially that of an unaffiliated critic. Bohr assisted other theoreticians in their calculations of the atomic model with orbiting electrons. But he soon realized that there was no way those models could possibly be correct. An electron moving in a circular path emits energy and would lose altitude. Ultimately, he concluded, any electron would crash into the nucleus. Inevitably. Conversely, given that matter is stable, the very fact that this does not happen was the real enigma. Bohr decided at some point during those months that the classical laws of electromagnetism evidently do not apply to electrons in atoms.

The exact day is not clear, but somewhere in 1912 Bohr realized for the first time that the solution to the problem lay in the very same quanta that had been so helpful for his theory of metals earlier. He designed a quantum theory for the very simplest atom, hydrogen, with a proton as its nucleus and a single electron orbiting it. The electron cannot adopt any old random orbit, but only those orbits that match the quantized energy of the electron. On top of that, the electron is able to jump from one orbit to the next as precisely the requisite amount of energy is absorbed or emitted.

In the first instance, there was hardly any reason to take Bohr's wild conjecture about the internal structure of the atom seriously.

And so the physicists didn't. They only started to get interested when Bohr linked his theory up to the light that hydrogen emits.

The spectrum of light that hydrogen atoms can emit, in sunlight for instance, contains a series of striking lines. The gas is clearly not able to emit any old amount of energy, but only specific values. The Swiss mathematician and physicist Johann Balmer had even derived a simple mathematical relationship for where those lines were positioned in the solar spectrum, without knowing where it came from.

However, Bohr's quantum-based atom explained Balmer's formula instantly. The spectral lines occur when the electrons in hydrogen atoms jump from one permissible orbit to another. When that happens, they only emit specific amounts of energy. Bohr demonstrated effortlessly why. It is that passage in particular in Bohr's articles from 1913 that makes his theory credible. And Bohr became famous, among his colleague physicists at any rate.

In a certain sense, Bohr was lucky. In the very same year, for instance, it turned out that his theory only applies precisely for hydrogen, the very simplest atom with a single proton as its nucleus and a single electron orbiting it. As soon as you get to helium, with two protons and two electrons, the calculations go pear-shaped.

Bohr, a restless soul who was always full of ideas and new approaches, kept fiddling with it, but his atomic model with the core and the electrons in orbits was simply too classical as well as being too primitive. Physicists came to realize that atoms are not miniature solar systems. They have their own laws and rules.

Quantum theory only really took off in the 1920s when a new generation of young physicists, including the *wunderkind* Werner Heisenberg in a starring role. Interestingly enough, they managed to do so by expressly jettisoning all their physical intuition and finding their way purely through mathematical relationships and smart calculations. It opened up a world of possibilities and mysteries that we will hear much more of later. However, the new wave of abstraction did not make quanta any less esoteric.

So that's a brief history of the early period of quantum physics, starting as a mathematical act of desperation and finally becoming the key to all the matter around us. From a historical perspective, it is not surprising that the discrete structure of energy, its fundamental graininess, initially appeared in the field of atoms and particles. That is the level at which events start to become pretty incomprehensible without the concept of quanta. At the very smallest level, the quantum dominates. The conceptual error that was then perpetrated for nearly a century is that real quantum phenomena can only play a role at the very smallest level, in cold and soundproofed laboratories, miles away from day-to-day reality. We will soon see that that is not the case, but it did mean that even physicists maintained that quantum theory was by definition exceptional, weird, and incomprehensible.

On the other hand, saying that it was a fatal flaw in their thinking would be going too far. After all, there was plenty of work for the new theoreticians in the world of matter itself. During the twentieth century, quantum theory triggered radically new and deep insights into both the chemistry and physics of solid materials in particular. And, although that term isn't used, each and every one of its applications could be called 'quantum technology'. Everything from protein chemistry to Intel chips.

I tap my iPhone recorder app to pause it. Time for more coffee. "What will it be, gentlemen? Anything with it, perhaps?" There's a waiter in the distance. I beckon him across.

2 PHYSICS FOR BOYS

IN WHICH WERNER HEISENBERG EXPRESSES QUANTUM PHYSICS IN NEW MATHEMATICS THAT SAYS PARTICLES ARE INTANGIBLE AND ELUSIVE, AND LOUIS DE BROGLIE SEES THAT PARTICLES ARE ALSO WAVES.

It's the same story every year, roughly when the poplar trees start to flower. Hay fever. While I am struggling to do what is absolutely necessary with my watery eyes and running nose, I often think back to a wonderful tale about the young German physicist Werner Heisenberg. Recovering on the German island of Helgoland from a bout of hay fever in 1925, he single-handedly changed physics forever. It is the romantic side of theoretical physics in all its glory, albeit a story that is bound to have been exaggerated a bit here and there. But that doesn't make it a tale that is any less worth telling.

I look up from my espresso in the bar of Hotel Métropole.

Bohr and Einstein nod in agreement. Of course! *Knabenphysik*, the 'physics for boys' from the 1920s, when quantum mechanics was truly born. The two of them were no longer boys by then, but they were part of it. Sure they were.

It couldn't have been any other way. The whole tale started with Bohr himself, after all, shortly after he put pen to paper at Rutherford's lab in England for his gutsy quantum theory of the atom. It was 1913 and the young Danish theoretician in Manchester had come up with the brilliant brainwave that an atom is a kind of solar system, with electrons orbiting around a central nucleus. He did not yet know exactly what those orbits were, but the crucial insight was that an electron can only move in orbits where the energy is a multiple of a basic unit of energy. He did not even know if that was the energy quantum that Einstein had suggested in 1905. But the model's approach worked well. When an electron jumps down from one orbit to another, a multiple of that energy unit is emitted in the form of light. Through a prism, that can be seen as a discrete spectral line. Bohr's theory was able to explain the basics of the spectrum of hydrogen; his sums matched up with the measurements that were available to within a few percent.

But the complications started immediately afterwards, pretty much while Bohr was still on his way home to Copenhagen. The light from elements other than hydrogen cannot be calculated at all using the same model. And even light from hydrogen that is passed through a good prism turns out to contain even finer details than Bohr's initial theory could explain. There are additional, vaguer lines that can be seen, and some lines are made up of two or more narrower lines. Physicists were faced with a conundrum. What was the message in the atomic spectrum?

In Germany, the theoretical physicist Arnold Sommerfeld took a pragmatic approach at first. Niels Bohr had introduced a number n *en passant* in his theory, the integer he used for numbering the atom's orbitals. It was in fact the first quantum number ever. But by no means the last. Sommerfeld, a professor in Munich, added two further quantum numbers, k and m, in an effort to fathom out the details of the spectrum. Various combinations of these were able to dictate the finer details of hydrogen light. It was unclear exactly what the numbers represented, but they provided some order in the chaos. At the same time, what had started as a description of the atom became an abstract numerical puzzle from then on. It would take until 1925 before it was resolved by one of theoretical physics' genuine sorcerer's apprentices, Werner Heisenberg, who had been taught by the very same Arnold Sommerfeld in Munich.

It was Heisenberg who had put up his hand after a lecture by Niels Bohr in Göttingen in 1922 and asked a critical question that the great man had no answer to either. Afterwards, Bohr invited the young student with the wide open, steely-blue eyes and spiky blond hair for a walk. The message was simple: you must come to my institute in Copenhagen some time. When he arrived in the Danish capital sometime around Easter 1924, the twenty-three year old German was impressed. Bohr's institute was bubbling with debates and brilliant young physicists who had all immersed themselves in the new atomic theory. Bohr's model had to be saved! Heisenberg soon became convinced that there wasn't much for him there. It wasn't helped much by the fact that he hardly ever got to speak to Bohr himself.

It stayed that way too. Bohr was a chronically overworked man in his thirties who operated without any obvious agenda or focus. He descended on anything and everything that crossed his path, which at the time hardly ever meant Heisenberg. Bohr's right-hand man, a Dutchman called Hendrik Kramers, did however immediately see the potential of the newly-arrived young German theoretician, and made time for him. During a second visit in the fall of 1924, he suggested to Heisenberg that it may simply not be possible to get a concrete picture of how the inside of an atom is structured. All we know about atoms, according to Kramer, is what they are able to tell us directly through the patterns of lines in their spectra. Perhaps the theoreticians should be concentrating on the observable variables, instead of the invisible inner workings of the atom.

Heisenberg started studying the patterns in the hydrogen spectrum, which evidently in some way reflected how the atom could transition from one energy state to another. Whether or not that actually meant electrons jumping across wasn't the point. He returned to Germany in April 1924, where he had been given a position with Max Born in Göttingen as a university physics lecturer.

At the end of May, Heisenberg was felled by a debilitating bout of hay fever. Until then, he'd been a healthy young man whose childhood had been full of nationalistic camping trips and who still enjoyed walks at weekends in the hills and mountains near Munich and Göttingen. He went to Helgoland, an island off the German North Sea coast, hoping that the clean sea air would bring some relief. When he got there, his eyes were nearly closed and his face so swollen that his landlady thought he had been beaten up.

As well as fresh seaside air, Helgoland let the young physicist rest and concentrate. Which worked wonders. Heisenberg tackled the weird energy jumps that an atom was evidently able to make. He imagined the atom as an abstract oscillator, a kind of system of springs that can vibrate in all sorts of different ways. He introduced two integer values into his calculations that enumerated the initial and final states of a transition in the oscillator. He tabulated them into rows and columns, a far from unfamiliar way for a theoretician to keep an overview of the relationships.

And he did just about manage to retain that overview. In one long night, Heisenberg worked out how the table of transitions results in a specific spectrum of light and checked that the whole thing complied with one of the strictest requirements in physics, conservation of energy. When dawn came, he closed his notebook and went to sit among the dunes. He had developed a new language for quantum theory, even if it was as yet too complex to understand properly. Later that day, he started writing an article about atomic states, defining the first rules of calculation for the new mathematical language. In Göttingen, his professor Max Born stared for quite some time at the lengthy mathematical expressions that Heisenberg had penned in his manuscript.

Then it clicked. Heisenberg's tortuous calculations were a clumsy version of something that any mathematician could handle with no problem: matrix multiplication. The new language of the atom is that of mathematics, with the states expressed as rows and columns of numbers, held together by other tables with yet more numbers. It was July 1925. The new field of quantum mechanics was born.

Once it had been cast into the new mathematical mold, the math immediately worked its own magic. The vectors and matrices that had evidently become the right tools for quantum theory possessed mathematical idiosyncrasies that got the physicists thinking. Heisenberg broke off the work at that point for a while. He first needed to reinforce his math skills and would always detest the much too mathematical name (or so he thought) that his theory would later acquire: matrix mechanics.

In the meantime, his teacher Born and his student Pascual Jordan discovered that the multiplication of some of Heisenberg's matrices wasn't something that could be done in any old order, in the way that you can with normal numbers. That was weird, because the matrices looked exactly as if they represented what classical physics would call a position, or a speed, or a momentum. The jargon term is that the matrices were non-commutative.

At the time, it wasn't clear exactly what that meant. But it was the first precursor to one of most widely known peculiarities of the quantum world: that it is not possible to know the position and speed of a particle exactly at the same time. The more precisely one variable is determined, the vaguer the other one becomes. And that has nothing whatsoever to do with careless or inaccurate measurements. The internal mathematical workings of the quantum world make it impossible for both variables to be precisely defined if both are requested at once. This is because speeds and positions are not mere numbers, as we are used to thinking of them in our large-scale world. They have a richer mathematical internal working.

We will later see that Heisenberg's uncertainty principle, as it became known, applies not only to the combination of speed and position but also for other mathematically related pairs of variables such as energy and time. One important thing to realize even at this stage is that the principle applies not only for physicists in their labs full of particles and atoms. Improbable as it may sound, Heisenberg's principle is for instance crucial for the way in which living tissues grow and repair themselves after being damaged. That's surprising, because the uncertainty principle is actually an almost stupidly small effect. For large objects with large masses and velocities, such as falling apples or racing cyclists, determining both speed and position is no problem at all. But for proteins and enzymes, for example, some theoreticians say that it is the only way that molecular barriers can be beaten or how bonds can be made that would be impossible from a classical point of view.

But let's get back to 1925, the breakthrough for boys' physics that emphatically no longer wanted to imagine atoms and particles as small balls zooming about. Although that yielded a theory that could tackle the issues, not everyone back then was equally happy with the complete abstraction that Heisenberg's method turned the atom and other quantum objects into. It also left open the question of what in God's name Planck's light quantum meant in the context of the internal workings of atoms, for example.

But even the physicists of the 1920s who preferred to keep the quantum domain more comprehensible ended up with insights in their models that could in no way be called 'everyday'. Take Prince Louis de Broglie, a French nobleman who had first read history and law before studying radio technology in the army, had slowly become interested in physics and a little later added a devastating new insight to that field. De Broglie was the one who somewhere around 1923 had already questioned the very principle of the difference between light and matter. Not because they are the same thing, but because they both have properties normally ascribed to particles or to waves only. Particles are waves and waves are particles. It depends primarily on what you're trying to find out about them.

De Broglie was fascinated by two famous formulae from the other guest at my table today, Albert Einstein. On the one hand, there is $E = mc^2$, the relationship between mass and energy that he derived in 1905 as part of the aftermath of his theory of special relativity. A certain amount of mass represented a certain amount of energy, and vice versa. The other relationship actually went right back to Max Planck's fortuitous discovery of the idea of packets of energy whose energy is related to a frequency. Planck saw that as the frequency of an imaginary oscillator, a spring-like system that represented the wall of a hot black box; the energy is proportional to the frequency, linked by Planck's famous constant h. Einstein had used that same relationship, also in 1905, to explain the photoelectric effect, in which light can release electrons from metal surfaces.

In a discussion with his brother Maurice, who ran his own X-ray lab in Paris, De Broglie wondered what would happen if you combined the two expressions. That appears to define a wave phenomenon – imaginary or otherwise – that matches the movement of an electron, for example, although an electron has mass. After all, a quantum of light at a given frequency does represent a certain amount of mass.

De Broglie's thoughts didn't actually go much further than that at the time. It was however immediately clear that any wavelike behavior of matter would be limited to small particles. In De Broglie's equation, the presumed wavelength of an object is inversely proportional to its mass. If the mass is large, the wavelength becomes immeasurably small. Bowling balls do not exhibit wave properties.

But electrons might. And if they did, it might be possible

to demonstrate that with measurements. Might it be possible to see a beam of electrons being diffracted, for example, like a beam of light? Could these waves of matter perhaps reinforce one another or cancel each other out, as happens when light waves interfere? It was all theoretical to the curious French prince, but it was a theory that got many experimenters' hearts beating faster. If electrons were waves, it ought not to have been all that difficult to demonstrate this.

De Broglie himself was more fascinated by the theoretical possibilities of his waves of matter, which even gave a relatively intuitive explanation for Bohr's atom. In Bohr's model, electrons are only able to go around the atomic nucleus in certain orbits, in such a way that their energy is a multiple of the basic energy packet, the quantum. It was clear to De Broglie what that meant in physical terms. If an electron is a wave, it can only go around a nucleus in a stable pattern if the wave fits exactly an integer number of times into one orbit, like a standing wave in a violin string. He published the idea in October 1923 in a French journal and it immediately drew a great deal of attention.

It was however hard to digest. De Broglie's courage was praised, for instance by the grand old man of Dutch physics, Hendrik Lorentz, and Einstein asked him for a copy of the article. But the idea of matter as waves also raised numerous questions. What is actually waving, for example? Einstein's theory of relativity had only recently freed light from the idea that some intangible medium, the ether, was required through which light waves could propagate. According to Einstein, electromagnetic fields could propagate themselves, even in a vacuum. Could something similar be thought up for these new waves of matter?

That wasn't what Einstein was most bothered about, though, in his subsequent response. He found the idea of stable standing waves of electrons around the atomic nucleus very palatable. That did however lead to a question of principle: why should electrons change orbits if the higher ones were just as stable as the lower ones? The idea of electron transitions in the atom perhaps being the result of a random process left Einstein with a nasty taste in the mouth. Not for the first time, or for the last. In 1920, he had already expressed his concerns to Max Born about the lack of genuine explanations in the up-and-coming quantum theory. In 1927, here in the Hotel Métropole in Brussels, he would keep expressing it succinctly to Bohr: *Der liebe Herrgott würfelt nicht*. God doesn't play dice.

De Broglie's waves of matter had a technical problem in addition to the issues of principle: a simple calculation showed that they propagated faster than the speed of light, violating the very principles of the theory of relativity. If something really was oscillating, then it couldn't be actual matter. In his articles, the Frenchman who thought it up spoke of a phase wave. Although to be honest, that didn't explain very much at the time. What is actually waving about? No idea.

But that was changed in one fell swoop when the Austrian theoretician Erwin Schrödinger published an article in January 1926 in a leading German journal that seemed to slot everything neatly into place. A few months earlier, he had held a lecture for his colleagues at the University at Zürich about the ideas of De Broglie about particles and waves. That included the idea that electrons can only orbit an atomic nucleus if their wavelength fitted an integer number of times into the circle.

The physicist Peter Debye (who was originally Dutch) was also there and afterwards he didn't mince his words, as usual. He said that the idea of electron waves was interesting, but the idea of waves of electrons fitting in neatly seemed too childish for words. Debye said that if something was oscillating, it ought to be described using wave equations. That was how grown-up physicists would do it.

Schrödinger didn't initially have an answer to that, but around Christmas that year he left for Davos in Switzerland together with his mistress for a brief vacation and to get away from his increasingly rocky marriage. The comment made by his colleague Debye kept running through his mind. If something was oscillating, it should be possible to describe it using wave equations.

At his holiday address, Schrödinger started fiddling about with a suitable wave equation. In classical mechanics, the problem of a system's energy state could be described as a mathematical manipulation of a function. Why, wondered Schrödinger, should that be any different in the quantum world? He wrote down an equation for an electron in an atom, underpinned by the sole condition that the momentum of the electron was related to a wavelength, exactly as De Broglie had proposed.

It went quickly from there. The solutions to the energy equation transpired to be waves that were always a summation of a small number of basic solutions. This summation was not complex: only two integers were needed. That was still abstract math. But when Schrödinger then calculated the energy solutions for an electron in a simple atom, it reproduced the atomic spectrum of hydrogen precisely. The lines in the spectrum depended on just two integer values, in exactly the same way. The mathematician and physicist Johann Balmer had already discovered that in the previous century. And Niels Bohr had invented two quantum numbers to get his own model of the atom up and running. Werner Heisenberg had turned it all round and treated the spectrum of light from hydrogen as an abstract numerical puzzle.

But for Schrödinger at the end of 1925, the quantum numbers simply dropped into his lap, despite the fact that he had stuck as close to classical physics is possible. So close indeed, that Heisenberg didn't trust it. His motto was that anyone who reckoned they understood quanta hadn't got the picture.

During my own student days, I had the Schrödinger equation on a large sheet of paper above my desk, and a significant proportion of the quantum theory exercises involved solving it. For a particle in a deep well. Or in a parabolic well. In a box. Up against a fixed wall. The procedure was always the same. Determine the boundary conditions, use a bit of math to solve the equation, then select the solutions that fit the conditions. Square the function you've found, and the places where that result is at its maximum are where you'd find the particle most often if you made a series of measurements. It became just another skill, but nobody seemed to be bothered how weird it was: particles that were actually abstract waves, yet particles as well. "Shut up and calculate," was the rule. Anything else was just food for philosophers, a profession many physicists have an innate dislike of because it's so much talking and so little calculation.

The bottom line is that the work done by Heisenberg, De Broglie, Schrödinger and many others in the 1920s meant that all the old conceptualizations of atoms and particles had become untenable. The picture of an atom as a miniature solar system – conceivable, albeit dominated by strange quantum rules – was finished. The only thing that we really know about atoms and matter is that they give off measurable signals, principally in the light that they emit. The new field of quantum mechanics was able to describe and explain the patterns in that light. But at a cost. Particles are not merely waves, but their positions and velocities are not hard-and-fast data either. And those waves aren't in fact the actual particles; rather, they are a yardstick for the probability that an attempt to make a measurement will find them at a particular place and in a particular state.

Vagueness and probabilities instead of concrete variables: none of it makes physics any simpler. The average observer would at most have been reassured by the assurances of quantum physicists that the weird phenomena that they were describing only play a role of any significance at the smallest scale. They were problems in the laboratory, not the real world, where the inherent vagueness and randomness of matter seems to have been smoothed away, overwhelmed, so small that it is perhaps negligible.

Quantum mechanics is therefore not only literally far removed from reality but also figuratively: quantum theory is strange, but otherwise harmless. It's the domain of a few physicists who want to take it further, plus a handful of aficionados of exotic and mind-boggling physics. Science fiction from a very distant planet.

That's a perspective that probably makes what follows all the more shocking. Quantum theory is by no means rarefied and distant. Numerous quantum effects come into play right under our very noses. To pick just one example, chemistry as a whole is in fact a fantastic illustration of the richness that just a few elementary particles and a bunch of quantum-based rules can generate. And that's just the start.

I can see Einstein and Bohr perking up. Ah, chemistry. Wolfgang Pauli. Samuel Goudsmit. George Uhlenbeck. Right: spin!

INTERMEZZO

CALLING THE BLUFF WITH TWO SLITS

IN WHICH THE STANDARD ISSUES IN QUANTUM MECHANICS ACTUALLY TURN OUT TO BE A SHREWD THOUGHT EXPERIMENT.

Now that we're here, drinking a lunchtime coffee with Einstein and Bohr at the Hotel Métropole in Brussels, there's something I have to confess. I've never felt comfortable, to put it mildly, with the books that taught me physics and the endless series of popular science books that followed. In my time, for quantum theory that meant Part III of Feynman's *Lectures on Physics*. The famous *Lectures on Physics*, I should add at this juncture, given that most physicists think that that's the right thing to say.

The American Richard Feynman was a giant, as I was told as a student by everyone, a Nobel Prize winner whose lectures were so amazing that they were all videoed, half a century before that became commonplace at universities. The *Lectures* were the paper version. Another aspect that played a role was that Feynman cultivated a rebel image, being photographed every now and then playing the bongos. Another reason why physicists loved him.

But I didn't, and in that context there was no more awkward issue than Feynman's explanation of the double-slit experiment. According to him, that famous experiment (there I go using that word again) exposes the very heart of quantum mechanics. Not only that, it shows what Feynman dramatically called "the most profound mystery of physics."

What that mystery involved was indeed instantly obvious. The double-slit experiment is a crystal-clear demonstration of the fact that thinking of particles such as electrons in classical terms is a pointless exercise. In essence, the experiment shows that an electron is a wave that can interfere with itself. More than that, even: in quantum reality, that's happening all the time.

The experimental setup is straightforward. A beam of electrons, similar to that from an old-fashioned television tube for example, is fired at a screen in which there are two parallel slits. A photographic plate or fluorescent surface has been placed behind that screen so that the impacts of the electrons can be seen, either directly or after developing the plate. What, we ask rhetorically, will be the pattern of intensities on the detector behind the screen?

Classical thinking says that there should be two lines, one directly aligned with each of the dual slits as the particles fly straight through them. However, in reality the back screen shows a pattern of more than two parallel stripes, a picture that any physicist would instantly recognize as an interference pattern. The conclusion is that a wave phenomenon must clearly be involved, even though we were using a beam of electrons. And that is of course exactly one of the key things in quantum mechanics: particles are also waves. Much as Louis de Broglie had in fact imagined, and much as Schrödinger's equations later showed so neatly. And as you yourself said in 1922, *Herr* Einstein, if wave properties can be ascribed to quantum particles, they should in principle be capable of mutual interference.

The most amazing part of Feynman's narrative about the dual slit experiment, though, is that the effect is not a collective one. Even if we reduce the intensity so far that the electrons are being fired at the screen one at a time, he tells us, an interference pattern will still appear on the second screen. The conclusion is that even a single electron does not pass through just one of the two slits. It goes through both.

Electrons are evidently waves and it is possible for the electron ripples to interact with each other. All it means is that it now takes longer to produce enough of a track on the photographic plate. If we watch live on a fluorescent screen, rather than using a photographic plate, we could even see the electrons impacting one by one. So those waves are in turn still particles.

In short, it's an intriguing and convincing experiment. And the result of the experiment does indeed lay bare a deep mystery: particles are waves, and waves in turn are particles. What then strikes me, however, is that Feynman says this is a thought experiment.

That starts to grate straight off. Can a thought experiment produce results that you haven't considered in advance? Fair enough if thinking exposes a paradox that makes you think again, or if a relationship appears that wasn't evident when you started. But the fact that an interference pattern appears on the photographic screen behind the two slits is neither of those. It's an experimental fact that you can at best verify.

Obviously, the reverse is also possible. If you consider that electrons are wave-like in their behavior, then you'd expect to see an interference pattern of light and dark strips on a photographic plate placed behind a pair of narrow slits. What Feynman essentially did was tell the story backwards. Undoubtedly with the very best didactic intentions, but as a student I felt that he hadn't been playing quite straight with me. The interference lines were a mystery, of course. But in fact, they were the result of a mystery that we ourselves had inserted in the first place, by assuming that electrons were also waves.

There had been plenty of reasons to do so, both theoretical and experimental, after half a century of quantum mechanics from Bohr and Einstein to Heisenberg and Schrödinger.

But how does that affect the real world? Weren't there experiments that would show an interference pattern? That would make Feynman's thought experiment a substantive, physical fact: if we fire electrons at a screen with two splits, we see such-and-such and that means that those electrons are exhibiting wave-like behavior.

Feynman believed, when he gave his lectures in 1961, that the dual-slit experiment was impossible to carry out in practice. But if it could be done, he knew that it would undoubtedly demonstrate the mysteriousness of the quantum world. It's a powerful tale nonetheless. Numerous books and articles about quantum theory use the parable of the electrons and the two splits as a demonstration of the inner mystery of the world of quanta. But that doesn't alter the fact that the whole double-slit experiment that you see in the textbooks was pure bluff.

Let's be fair, back in 1961 Feynman was in formal terms both correct and incorrect. He was right when he reckoned that the dual-slit experiment was awkward to carry out in practice. But wrong in the sense that it really had already been done. At the end of the fifties, the German physicist Claus Jönsson did an experiment with the slits at the University of Tübingen (using five rather than two, for technical reasons) and observed exactly what the theoreticians had expected: a pattern of peaks and troughs in the intensity at points behind the slits where the electrons did and did not hit the plate. Jönsson published his work in his thesis in 1961, so Feynman could therefore have known about it. But the thesis was in German. And in fact, physicists had known for certain for some time that electrons acted as waves when appropriate. They didn't need Claus Jönsson or the two slits for that at all.

In 2002, the British journal *Physics World* carried out a survey to see what people thought was the greatest experiment ever in physics. Readers submitted more than three hundred suggestions, with the double-slit experiment as a very clear favorite. Except that nobody mentioned the name of Jönsson. The experiment with the electron and slits was ascribed by most of the respondents to Thomas Young, although he had actually used two slits with a screen behind at the beginning of the nineteenth century to demonstrate the wave nature of light. The interference patterns were virtually identical to the ripple patterns behind a quay with two openings that waves were sloshing up against outside: those two apertures in turn become sources themselves, generating circular waves that can reinforce one another or cancel each other out at various points further along.

In historical terms, the confusion isn't so surprising. The wave nature of electrons had in fact already been demonstrated in 1927 by firing a beam of electrons at a nickel crystal (C. Davisson and L. Germer) or celluloid (G. P. Thomson) and seeing where they then ended up. Every aspect of the patterns this produced suggested interference between waves. They look just like the patterns of light that Thomas Young had used to demonstrate the wave nature of light, for example.

When a German doctoral student carried it out for real somewhere around 1959, the double-slit experiment was basically just a bit of didactic legerdemain, meant to amaze the student or reader and drive home to them just how very peculiar quantum mechanics could be. The trick works and so it's no surprise that plenty of books still discuss it in detail. But ultimately, it's the conjurer's magic rather than the mystery of the quantum itself that amazes. Whereas real physics starts with facts.

And Feynman? I did start to appreciate Feynman properly much later. He did some wonderful physics, so original that it is often difficult to fit it in with everything else, and he also wrote some fantastic books. His memoirs *Surely You're Joking, Mr. Feynman!* and *Tuva or Bust!* were deservedly bestsellers, not dissimilar for a young physicist to what rock-'n'-roll is for music lovers. The simple way in which he explained in the summer of 1986 that the Challenger space shuttle had blown up that January with seven astronauts on board because of a leaking O-ring is still legendary.

But despite all the bravado, he could also be an extraordinarily profound physicist, particularly in his later years. There are moments when I just simply have to browse through *The Meaning of It All* or *The Pleasure of Finding Things*

Out before going to sleep. Both of these are primarily an unadorned anthem to the amazement and wonder that Feynman believes is the driving force behind all good science.

In that light, the dual-slit experiment – whether or not it was imaginary – is pretty much the best example that physics has to offer. Even if it is generally told backwards.

"A pity that we were already dead by then," say Einstein and Bohr at much the same time. "We'd have liked to hear him tell it."

3 QUANTA IN A TEST TUBE

WHY ATOMS ARE NOT SPHERES BUT DO HAVE SHAPES, AND WHY ELECTRONS DON'T LIKE BEING STACKED TOGETHER – SO THAT THERE'D BE NO SUCH THING AS CHEMISTRY WITHOUT QUANTUM MECHANICS.

On a busy roundabout in Rotterdam, 's-Gravenplein to be precise, there is a gigantic statue in front of a former school building. Worthy of a king, it's probably the biggest statue in the Netherlands of a scientist. The man on the verdigris-coated bronze pedestal high above the viewer is Jacobus van 't Hoff from Rotterdam, who was a professor in Amsterdam and most notably the first Dutch Nobel Prize winner. That was in 1901.

More than a century later, that prize still seems fully justified. Van 't Hoff, who incidentally worked in Berlin for much of his career, is seen as one of the founding fathers of modern chemistry. He wasn't particularly into laboratory experiments – he was both too clumsy and too impatient for

that. A bit of a slob, really. But Van 't Hoff was a brilliant thinker. In 1874, a few months after his not particularly impressive thesis, he wrote a pamphlet about a dozen pages long that would make his name. Molecules, he stated in it, have a spatial structure.

They can have a front and a back, and more importantly some molecules can appear in specific forms that are mirror images of each other, which can also behave differently chemically. Lactic acid for instance has tetrahedral geometry, a triangular pyramid around a central carbon atom C, with different groups such as H, OH, COOH, and CH_3 at the vertices. Those groups can however be arranged at the corners in two genuinely different ways. Lactic acid therefore occurs in two forms, which can for instance be distinguished by the fact that they scatter light in different ways. That was already known in Van 't Hoff's time, but he was the first person to think of an explanation. In modern-day pharmaceuticals, this type of chirality plays an essential role. Because – perhaps surprisingly – life itself turns out to have a huge preference for one type rather than the other.

Brilliant or not, the daydreaming and messy Rotterdam doctor's son (and chemist) Van 't Hoff searched in vain for a job at first. At one of his many job interviews at secondary schools in Breda, Dordrecht, and Leeuwarden, he enthusiastically took his colored cardboard models of molecules from his pocket, models he liked to use when he met people to illustrate his radical way of thinking. The interview committee didn't understand a great deal of it and played safe by not giving him the job. The statue that has stood in front of the entrance to that very same school ever since 1915 can be seen as a quiet triumph.

Van 't Hoff's original pocket cardboard models are currently kept at the excellent Boerhaave Museum in Leiden. In the same room, the gentlemen here at my coffee table in Brussels will be pleased to hear, where one of Einstein's own fountain pens is exhibited like a holy relic along with various other things left behind on his numerous visits to that important city.

Unlike the ball-and-stick chemical models of salts and molecules that I myself got to use during my schooldays, Van 't Hoff's molecular models were rather abstract, constructed of red and yellow painted cubes, pyramids and tetrahedra. He was not yet using anything like balls for the atoms and sticks for the forces between them, which is of course not so strange. When he was starting to think about the spatial shapes of molecules in around 1870, even the majority of chemists didn't really believe in the reality of atoms and molecules. Even they tended to see these kinds of representations more as theoretical tools, merely intended to help you understand the invisible inner workings of matter.

Ultimately, it would take until the 1920s not only before the reality of atoms was irrefutably demonstrated, but also before science began to realize why atoms and molecules really did have spatial shapes. It didn't just happen. It needed an Einstein, who in 1905 was the first to work out the actual dimensions of molecules. And a Bohr, who in 1913 saw an atom as a quantum system, and a Sommerfeld who wrote down the basic rules for it. And Heisenberg and Schrödinger, who somewhere around 1925 envisaged the internal machinery of the quantum atom, one of them using pure math and the other using more easily pictured electron waves.

But above all it needed a Wolfgang Pauli, whose work

with others such as the young Dutchmen Samuel Goudsmit and George Uhlenbeck in Leiden added something to quantum theory that had always been missing until then: preferential directions. The story starts with the electron, which turned out in the 1920s to have not only a charge and a mass, but also what was referred to as a spin. This atomic building block appeared to act like a spinning top, going clockwise or anticlockwise. Although you do have to understand of course that nothing is quite what it seems in the world of quanta.

The story of spin can be introduced in a variety of ways, but the simplest starting point is perhaps a question. Why don't atoms collapse? Niels Bohr had used his quantum models of the atom to show why atoms only radiate energy in very specific packets, as well as demonstrating that hydrogen stuck neatly to the rules he had thought up. Until then, the fact that it didn't work very well anymore for the rather larger helium atom (two protons in the nucleus, two electrons going round it) seemed only to be a question of corrections. But there was a more fundamental problem in larger atoms. You would expect all electrons to drop naturally into the lowest energy level, but there was clearly something preventing that from happening. Electrons in carbon, for instance, are not all in the lowest orbits. On the contrary, that can never happen. So what's stopping them?

As was often the case early in the history of quantum mechanics, the solution was initially more of a guess than a genuine derivation. The key player in this case is Wolfgang Pauli, an Austrian *wunderkind* who joined Sommerfeld in Munich in 1918. He was then aged just 18, but full of confidence. When the great Niels Bohr was giving a guest lecture to the group, the youngster Pauli spoke up during a round of questions, saying that Mr. Bohr had certainly not been saying anything silly, but there were a few points... Bohr immediately invited Pauli to come and work with him in Copenhagen. Like that boy Heisenberg, this young Wolfgang was a talent that he wanted to encourage.

Pauli came to Denmark in 1923 and he did not disappoint. In Copenhagen, he tackled the problem of the structure of the atom, for which a confusing plethora of new ideas were doing the rounds at the time. Many of them were entirely *ad hoc* and seemed little more than cabbalistic, esoteric mumbo-jumbo. In the meantime, many physicists had become convinced that electrons within atoms were present in layers or 'shells'. But exactly how many there were and in what shells was still open to debate. Pauli discovered that the disputes basically boiled down to a factor of two. Each shell seemed able to accommodate precisely twice as many electrons as expected, and Pauli – young and undaunted – did not hesitate: electrons that otherwise seemed to be identical in all other ways must be distinct in terms of some unknown property. The patterns in the periodic table of the chemical elements suggested furthermore that there must be some kind of prohibition on having absolutely identical electrons in the same shell. This let Pauli calculate that two electrons fitted in the first shell, eight in the second, eighteen in the third. That is a pattern that chemists would recognize instantly as the numbers of chemical elements that occur between the successive noble gases in the periodic system.

In December 1924, Pauli described his exclusion principle in an article for Bohr in Copenhagen. Bohr was astonished by just how effective and simple the idea was, but also immediately asked the crucial question: in what way can these otherwise identical electrons be essentially different?

The answer to that conundrum came from two young physicists from Leiden, albeit originally without them realizing it fully. Samuel Goudsmit, or Sem to his friends, was one of Professor Paul Ehrenfest's doctoral students. Ehrenfest had been keeping an eye on the young Dutchman, who had followed up his studies in Leiden with a period in Italy as a private tutor at the Dutch embassy, but now wanted to study with the professor for a doctorate. George Uhlenbeck was a top-class theoretician but he had missed a lot of the latest developments in quantum thinking. And so Goudsmit, who was by then already working three days a week as an assistant theoretician with Pieter Zeeman in Amsterdam, was given the task of bringing Uhlenbeck up to speed on that particular field in short order. The two became firm friends, and were indeed to cross The Pond a couple of years later, to work at Ann Arbor in the United States. Every Wednesday, they first attended Ehrenfest's famous colloquium at his villa on Witte Rozenstraat in Leiden. Then they went painstakingly through the new physics together, in the process coming across Pauli's peculiar principle and the doubling up of electron numbers.

In his memoirs, Goudsmit described how Uhlenbeck – not prejudiced by any pre-existing knowledge of the subject – instantly saw what was happening: the electron must have some extra property, which they were able to envisage as an internal spin. Goudsmit and Uhlenbeck started calculating. A rotating charge has a magnetic orientation. If an atom is placed in a magnetic field, the internal magnet could be opposed to it or aligned with it. That gives two slightly different energy levels, meaning that it should be possible to split the lines in an atomic spectrum into doublet pairs using a magnetic field. That is the effect that Goudsmit's boss Pieter Zeeman had become world famous for and that led to him getting the Nobel Prize.

However, at that time another interpretation held sway for the Zeeman effect and a number of subtle variants on it, conceived by the German theoretician Alfred Landé. He suspected that the extra magnetic effects arose because electrical charge is moving around within the atom, in the form of electrons. That supposedly created a current loop that produced a magnetic field. But there was a problem with that, in that the magnetic field could (according to the classical viewpoint) be oriented in any direction. In Bohr and Sommerfeld's quantum theory, though, only specific directions are permissible. To make things worse, a famous experiment by Otto Stern and Walther Gerlach had just shown that silver atoms genuinely did have their own magnetic fields, but in just two directions. There was clearly something wrong with the idea that electrons whizzing about could be responsible for the behavior of atoms in magnetic fields.

Uhlenbeck and Goudsmit realized what the issue was. The magnetic effect everyone was looking for had nothing to do with the orbits of the electrons, as Landé believed, but was in the electrons themselves. Uhlenbeck summarized the results in a brief article in both their names and sent it to Lorentz, asking him to submit it to *Naturwissenschaften*, one of the leading journals of the time. That didn't go quite as expected. Lorentz took a look at the paper and saw a problem: for the given values of the spin, the surface of the rotating electron would be moving much faster than the speed of light. That was a real difficulty, actually requiring a relativistic treatment of the entire question.

The skeptical answer bothered Uhlenbeck and he decided that the article had evidently not been thought through properly. He told Lorentz that he was withdrawing it, but it was too late by then. Despite his reservations, Lorentz had submitted it for publication and the article had already been accepted. The article by Goudsmit and Uhlenbeck appeared at the end of 1925 and the two young physicists from Leiden were suddenly sought-after partners for the big guns in the field, from Einstein to Bohr.

Two years later, they received their doctorates on the same day for related topics in atomic spectroscopy. A short while after that, they and their young wives embarked on a ship for Michigan, where Goudsmit would make his name not only as a professor but also as the editor-in-chief of *Physical Review*, one of the most important journals in physics. In 1958 he thought up the *Physical Review Letters*, which with its rapidly published articles is seen as the most authoritative physics journal in the world even in our era of online publications. Uhlenbeck returned to the Netherlands in 1935 and obtained a professorship in Utrecht, but left for the USA again in 1939, this time to escape the burgeoning anti-Semitism. He was only just in time.

There is also a more tragic tale associated with the spin of the electron. A young German physicist by the name of Ralph Kronig, who had obtained his doctorate in New York, arrived back in his native country in January 1925. He had been thinking for some time that the supposed double-valued nature of the electron within the atom could boil down to two different ways in which the electron could spin about its axis, to the left or right. A rotating electric charge becomes a small magnet, with the direction determining how much energy it has when it is placed in another magnetic field. This could for instance explain why some spectral lines from various atoms split up into doublets when in a magnetic field.

When he met Pauli, though, the man was at his most fearsomely and bluntly cynical. He called Kronig's intrinsic rotation an amusing flight of fancy, *ein witziger Einfall*, but said he didn't believe any of it and didn't think that it really helped theory to progress. Kramers and Heisenberg didn't buy it either. Discouraged, the rather timid Kronig put the idea of intrinsic rotation to one side. Barely a year later, Goudsmit and Uhlenbeck published an idea that at first sight looked much the same, and which led to discussions among physicists and was ultimately embraced. The electron is not a rotating electrical charge, but it does have an inherent magnetic direction. Paul Dirac in England christened it the 'spin' and showed that otherwise identical electrons can only appear in the same atomic shell if they have opposite spins. Pauli did apologize to Kronig years later for his overly hasty rejection in 1925, which he himself ascribed to youthful overconfidence. Which makes it all the more painful that Pauli was awarded a Nobel Prize in 1945 for his exclusion principle. Kronig, Goudsmit, and Uhlenbeck would have been logical choices too, he acknowledged.

But that's not how it went, and in fact a Nobel Prize has rarely been so justified as the one Pauli received. Not only does his exclusion principle provide a superb pathway through the periodic system of chemical elements, but in fact chemistry as a whole – including the spatial nature that Van 't Hoff had discovered back in 1875 and won a later Nobel Prize for – is deeply rooted in it. In a chemical bond between two atoms, both parts are involved in a subtle quantum interplay with a number of shared electrons in the outermost shells. When those shells are not neatly radially symmetric but instead have certain preferential directions as indicated by quantum theory, bonds are therefore created at specific angles. This naturally creates structures that are three-dimensional rather than flat. And that includes the possibility of molecules that are chemically identical, but spatially different.

Van 't Hoff never got to know it; the quantum theory of atoms and molecules came about half a century after he had tried to explain the principle of stereochemistry to anyone who cared to listen, using his little cardboard models. But he was certain that spatial structures played a role in chemistry. Why not? Nowadays it tends to be the other way round. Chemists use quantum theory, generally on a computer, to predict the structures of previously unknown molecules, in particular for proteins and other biomolecules. Only then do tests in the lab follow, to see if the quantum chemists got it right.

"*Und ach ja, die Sem,*" says Einstein at our hotel table in Brussels as the setting sun colors the sky outside red. When Wolfgang Pauli won the Nobel Prize for physics in 1945, he would naturally have thought of Goudsmit and Uhlenbeck. Without their intuition about electron spin, the statistics of electrons in the atom might have remained a puzzle for a long time. And maybe the two lads from Leiden might also have been awarded a share in Wolfgang's prize money.

Although Sem also had other things on his mind, of course. After all, hadn't he been one of the people who had been with the advancing Allies, looking for Hitler's atom bomb research in the ruins of Germany? After what had happened to his parents too, carted off to Auschwitz while he was still trying to get them out to the States. Awful. He wrote about it touchingly in his book, after the war. How he had stood in his parents' plundered and dismantled house in The Hague, picking his old school exercise books out of the rubbish in the garden. *Armer Kerl* – poor guy. That feeling of guilt. Never quite the same again.

4 ALL ELECTRONS COUNT

HOW ELECTRONS IN SOLID MATTER FOLLOW THEIR OWN QUANTUM RULES, ALLOWING MATERIALS TO BECOME SEMICONDUCTORS THAT ARE SUITABLE FOR MAKING TRANSISTORS AND SILICON CHIPS. GADGETS USE REAL QUANTA.

The screen of my iPhone lights up for an incoming call and draws the attention of my intellectual table guests. I get rid of the caller, but it's too late. "I'd already been looking at that little device," says Einstein, something plainly telling him that he and his colleague Bohr have more to do with it than they could ever have guessed at the time, long before their deaths. Even though, he suspects, the route may have been somewhat tortuous.

Although he himself had messed around with friends such as Habicht during his time in Berne on some of their own minor inventions (what would you expect from a third-grade patents clerk?), he had never really had much to do

with electronics. He has to admit, doing his own experiments wasn't really in his nature. Pen and paper are simply, well, less impatient.

"Fair enough," I agree. "There aren't many labs that are more powerful than your own brain, Professor. If there's one person who's perfected the thought experiment, the *Gedankenexperiment*, then you're it. Your imaginary journey riding the crest of a light wave was once the very seed of the whole theory of relativity. Anyway, it's not as if you were all that impractical."

I tell him something that I've spontaneously remembered: the Boerhaave Museum in Leiden still has an example of one of his devices, designed for measuring extremely weak currents and once purchased by Pieter Zeeman's lab in Amsterdam. He grins. "*Ach so. Das Machienchen.*"

Niels Bohr mutters something about getting back on track and lights his pipe, the Brussels brasserie's ban on smoking notwithstanding. Nobody around us seems to notice.

Before we can describe exactly what an iPhone has got to do with my table guests Bohr and Einstein and their quantum theory, we need to take quite a detour looking at the power of statistics. The two of them had plenty to do with it, let's make that clear straight away. But first, a little diversion.

This detour starts with the inherent angular momentum of the electron, a kind of internal rotation that isn't really a spin at all but essentially an additional quantum mechanical property that can have one of two values, 'up' or 'down'. Pauli used atomic spectra and other experimental data to derive the fact that the electrons in atoms are distributed across a number of shells, with no two electrons being allowed to exist in exactly the same quantum state (including the spin). As a result, the shells in an atom are filled up in a very special sequence: two electrons fit in the first one, one spin-up and the other spin-down, the second shell can contain eight electrons and the third can have eighteen. Back then in 1925 it was not clear exactly what the spin was, but it was instantly clear that it played a key role in the structure of matter. The extra property organizes the electrons in a material in specific ways, so that they sometimes make it an insulator and sometimes a conductor. And sometimes they can also be what is known as a semiconductor.

It first became clear that quantum properties can dictate how particles form a substance and the properties it will have when Albert Einstein received a letter in 1924 from the unknown physicist Satyendra Nath Bose at the University of Dhaka (then in India, now in Bangladesh). An article in English was enclosed about the radiation of a black body, the very question that had led Max Planck to posit energy in packets – quanta – for the first time a quarter of a century earlier. Bose asked Einstein to translate the piece into German and submit it to the prestigious journal *Zeitschrift für Physik*. If he thought it was worth it, at any rate.

"The rest is *Geschichte*," concurs Einstein contentedly from the other side of the coffee table. "History." He did the translation and published the article, adding a footnote of his own stating that he believed that Bose's statistical technique for an ideal gas of light particles – photons – also applied to a gas of material particles such as atoms. In the months that followed, Einstein developed the ideas further in a series of lectures for his fellow professionals in Berlin. On the other side of the world, Bose had no idea of the wheels he had set in motion.

The problem that first Bose and then Einstein was tackling was how the energy that radiated from an oven can be made up of the energy of photons if they are treated as a gas. Attempts to do that using classical theory never got anywhere. But Bose decided that photons are not distinguishable from one another and that any imaginary interchanges between them cannot have any influence whatsoever on the energy. That condition led to a new kind of energy statistics that would later become known as Bose-Einstein statistics.

Those statistics turned out to the physics' equivalent of Pandora's notorious box. Einstein followed the calculations through and discovered that an ideal quantum gas would have a specific critical temperature below which the particles in the gas would come together in the same state. In such a Bose-Einstein condensate, the particles lose their individual identity and there would then be no forces acting between them. The effect was first observed for real in 1995 by researchers in Boulder, Colorado, in a cloud of extremely cold rubidium atoms.

And much earlier, back in 1927, Willem Keesom discovered in Leiden – using a laboratory full of cooling equipment that he had taken over from the late Heike Kamerlingh Onnes – that liquid helium loses all its viscosity at a critical temperature of less than two degrees above absolute zero. That kind of superfluidity was also something that Einstein and Bose had foreseen in their quantum gas theory.

Whereas Bose-Einstein statistics for particles primarily has a role to play at the extremely cold end of physics, statistics turn out once again to dictate how matter behaves at the other end of the scale. But it is then a different kind of statistics. The key difference is spin. Bose-Einstein statistics only apply for particles with no spin (such as photons, which are mutually entirely interchangeable) or integer spins (such as many atoms). As soon as half-integer spins are involved, as in the case of electrons, it becomes more complicated because there can always be two versions of every quantum state, one with the spin up and the other with the spin down, giving them slightly different energies from one another. That leads to another type of result for the gas energy, described by what are known as Fermi-Dirac statistics.

That led to particles that behave according to this set of rules becoming known as fermions, after the Italian physicist Enrico Fermi, another *wunderkind* who made his mark on physics during the first half of the twentieth century. And indeed on the history of the world. He was the one who first succeeded in creating a nuclear fission reaction in uranium, in Chicago in 1942. The experiment was carried out underneath the concrete stands of the local baseball stadium, Stagg Field. If one of his assistants had not made a timely intervention by throwing a carbon rod into the improvised reactor that had started up the chain reaction, Fermi's actions might have caused a disaster in Chicago.

But that was later, in 1942. Sixteen years earlier, Fermi and his British colleague Paul Dirac – both brilliant but unconventional theoreticians – had been trying to work out what Bose-Einstein statistics would look like for particles with half-integer spin. Pauli's exclusion principle does not apply to bosons such as photons. But it does to electrons. What would a gas made of that type of particle do? As a rule, it would create a kind of pool of particles in which the energies could be substantially different. At the same time, there are no discrete energy levels in the pool: the electrons in it are interchangeable without very much happening. The calculations done by Fermi and Dirac also showed that there are sometimes energy levels above that pool into which it could be possible to raise an electron. That shouldn't be taken too literally, by the way. The electron remains in the pool, but its energy is essentially different from the others. It has only been 'raised' in terms of the energy diagram. All electrons count, of course, so at the same time this has left a hole in the pool.

This description was particularly an eye-opener in the understanding of solid substances. The image of a pool of electrons, referred to by physicists as the valence band, suddenly explained why some materials such as metals are conductors, whereas others are not. And why some materials haven't quite worked out which they want to do – conduct or insulate.

Only the very outermost electrons of the atoms are important for the chemical bonds in a substance. When atoms stick together, the outermost electrons are the ones that get more freedom of movement, essentially as determined by Fermi-Dirac statistics. The electrons in a substance that are more or less free remain in the lowest energy pool. The electron level then determines the properties of the material, in particular the electrical ones.

In a metal, the pool of electrons – the valence band – is not completely full. This means that electrons have countless opportunities to move through the material, without much energy being required. As soon as an electrical voltage is applied across the two ends, they start moving. A current flows and its strength depends not only on the voltage but also on the electrical resistance of the metal. In this case, the valence band is referred to as the conduction band.

And what about insulators? If the valence band in a certain material is completely filled, the situation is different. Even if an electrical voltage is applied, the electrons in the material are still unable to move. All the possible options have already been used up. There is no conduction band. Electrons would only be able to start moving via a different energy level, higher up. In insulators, they simply don't have enough of a push, or the jump is simply too high for them.

But there are also intermediate cases. In silicon, for example, the lowest electron pool is indeed full. But the next energy level up is not so very far above it, meaning that some electrons that have just a bit more zip can in fact jump spontaneously into it. And at that energy level, they are free to move. In particular when an electrical voltage is applied, a certain current will start to flow in what is called the new conduction band. At the same time, a gap has appeared in the lowest electron pool, like a positively charged hole, and it is able to move under the influence of that same voltage, but in the opposite direction to the electron that made the jump.

The discovery of Fermi-Dirac statistics – or perhaps the formulation would be a better word – was the precursor to a revolution in the physics of solid materials. Suddenly they were no longer stacks of atoms in salts or other compounds, they were quantum systems with properties that were dictated by the way in which electrons (in particular) have to arrange themselves in order to meet all the rules imposed upon them. That was not only interesting to study, but it quickly transpired that the new insights could also lead to wonderful new applications. These applications are virtually infinite in number, but there's no doubt that the key one is electronics. Without a proper theory of electrons, in semiconductors in particular, all kinds of things from modern smartphones to advanced solar cells would never have been possible. And the theory in question is of course quantum theory, or at any rate derived from it.

Semiconductors had in fact been known since the mid-nineteenth century, largely for the property of being able to conduct an electric current when light falls on them, but otherwise hardly at all. Major figures such as Faraday and Becquerel experimented with them. Crystals of silicon and germanium behave like that, for instance, as do gallium arsenide and selenium. As long ago as 1880, Alexander Graham Bell used the last of those materials to transmit sound vibrations via a light signal and then make it audible again. They were used for decades in the twentieth century to make light meters for photography. Silicon is one of the standard materials in the current solar cell boom.

Natural semiconductors, however, turn out to be just the tip of a massive iceberg of physical possibilities. The precise location of the valence and conduction bands in a material can be manipulated accurately by adding small amounts of impurities of other elements, which is known as 'doping' the material. Traces of phosphorus do this in silicon, for example. In 1940, the physicist Russell Ohl, scion of an immigrant family with Dutch origins and employed at AT&T, made the first measurements of doped silicon.

Doping can make it harder or easier for electrons to jump from one band to the next, meaning that the sensitivity to electrical voltages can be tweaked or even that sensitivity to light can be introduced. During the first half of the twentieth century, physicists and chemists started experimenting with this, at first largely by trial and error, but then gradually more and more led by the theory. Discoveries slowly merged into inventions, and engineers started to get involved.

With unprecedented success. In December 1947, a few days before Christmas, the physicists Walter Brattain and John Bardeen of Bell Labs demonstrated an inconspicuous little device to their bosses for the first time. It was no bigger than the nail of your little finger, a fragile construction made of a copper sheet, a crystal that they told the managers was germanium, a plastic wedge with gold foil on, held in place by a paperclip. Connecting wires protruded from it here and there, going to an ammeter with a glowing green screen. Every time Brattain pressed a button, a flickering spot of light hopped up by one grid square, showing that the output voltage was a couple of times higher than the input. While the snow was piling up outside and most of the staff had already headed homewards for safety's sake, the men inside had just witnessed the world's first semiconductor transistor.

The world would never be the same again. Until then, amplification in electronics had to be done using valves – vacuum tubes that are heavy and fragile and get hot (and therefore gobble up lots of energy). People had been looking for a solid-state alternative since the thirties. Everyone knew it would be based on semiconductors, but the question was how.

In September 1948, the journal *Electronics* put the discovery on what has since then been a contentious front page. The photo shows the group leader, William Shockley, behind a microscope as he manipulates the contact wires to an invisible transistor in front of him. Behind him stands Walter Brattain, hands on the back of Shockley's chair. John Bardeen watches from the side, his right hand holding a fountain pen and resting on a notebook. The photo may suggest harmony, but the real history of the discovery of the semiconductor transistor is not so uplifting. Brattain and Bardeen were the first to ask for a patent on their invention, but Shockley blocked it six months later, officially because of a technical error.

A transistor is not complicated. A voltage is used to create a current in a piece of semiconductor material, with the strength of that current depending on the voltage applied to a third contact. That control voltage changes the location and population of the valence and conduction bands of the semiconductor and therefore also its electrical conductivity. A small signal applied to the third terminal, such as a waveform in the voltage, can thus be amplified as much as you want in the output current. That offers some wonderful new opportunities for designing electronic circuits, making them ridiculously small by the standards of time. During the sixties, the term 'transistor' was synonymous for a while with the small portable, battery-powered radios that replaced the old valve-driven beasts with those glowing eyes and a power cable. Nowadays, semiconductor circuits are integrated by the billions at a time onto silicon chips. The details are so minuscule that short-wavelength UV light is needed merely to get them in the right place. Everything from our phones to our tablets and laptops use them, both in data processing and in the microscopic LEDs in full-color screens. The word 'transistor' has already been almost forgotten. Virtually nobody knows that the quantum statistics of semiconductors played a crucial role. The element silicon also gives its name to a high-tech valley in California, the birthplace of all modern digital technology, from hardware to software.

Bohr weighs the smartphone approvingly in his hand and then puts it back down on our table. "Nice. And it's amazing what the old quantum theory has apparently led to," he says. "How tangible it's actually all become. And that name too, Silicon Valley. Marvelous." He would like to have seen more of that during his own lifetime.

"Your words suggest," he continues, "that this device is actually just one modest example of how quantum theory has been utilized?"

The two old physicists look at me expectantly. "You can count on it," I begin. But that small pun escapes them.

INTERMEZZO

MAJORANA GHOSTS

HOW A QUANTUM PARTICLE APPEARED ON A CHIP IN DELFT, AFTER A WAIT OF SEVENTY YEARS.

Somewhere in the early spring of 2011, I'm sitting in the office of the quantum physicist Leo Kouwenhoven in Delft. On a virtually empty desk there is one lonely PC. It's switched off. As well as years of scientific journals, theses, and books, the tall cupboards lining the walls are also full of the next generation's colorful scribblings. The winter sun is shining outside. Our espressos are starting to get cold.

I happened to be in the area, so we were discussing our fantasy of writing a book together for the umpteenth time. A book about the misunderstanding that quantum mechanics is something exceptional, a world apart. A weird and wonderful world at that, one that ordinary people are incapable of understanding anything about. One that even the majority of physicists say, for the sake of simplicity, that you shouldn't even try to understand. Knowing the quantum laws and being able to work with them: that's something for physicists.

According to Kouwenhoven, that is both a misunderstanding and taking the easy way out. His entire research career is focused on understanding quantum effects in the reality around us. Which means that quantum reality can't be all that peculiar. Our book, I have already suggested, should therefore be called *Real Quanta*. Because it really will be about quanta in the real world, as well as how quanta are more real than we usually think.

Leo and I have known each other for years. He is the physicist who, in an icy cold Delft somewhere around Christmas many years earlier, was ready to show a lonely postdoctoral student like me how miniscule pyramids of semiconductor could suddenly start exhibiting atom-like properties. And now, as a science journalist for a newspaper, I'm always looking for exciting news. The meeting led to a nice little story about so-called quantum dots. Nice for the cognoscenti any rate. There's a kind of alchemy going on there in Delft. They use vaporization and etching to kludge together artificial atoms that nature forgot to make.

And Kouwenhoven has some exciting news this time too, though I most emphatically still have to keep it under my hat: the software giant Microsoft is going to invest a million euros in work that he has actually largely had sitting there on the shelf for some time, because the Americans think it might yield the key to a working quantum computer. The only thing the Delft physicists have to do is show that the Majorana particle really exists.

That catches me for a moment. *What* does he have to show? Sure, I know who Majorana was. Ettore Majorana. An Italian physicist, a theoretician from Enrico Fermi's entourage, who put a number of fantastic concepts on paper, about spooky neutrinos for instance, particles that are produced by nuclear reactions, fly straight through everything, have virtually no mass and – who knows? – could even be their own antiparticles. And Majorana is also the physicist who disappeared at a young age in 1938, purportedly during a trip to Sicily. There were rumors about depression and suicide. There was even speculation about an abduction to Russia. The reality is that nobody knows.

Young Italian physicists nowadays all still know who Majorana was. Me too. An Italian once told me an undoubtedly exaggerated version of this tale when I visited the Fermilab accelerator laboratory in Illinois, where he worked. But don't ask me exactly what this somber genius of a theoretician Majorana actually thought up...

Leo helps me out. Majorana is the guy with the particle that is its own antiparticle. Around 1935, he derived from his equations for quantum particles that something like that could exist, and perhaps even ought to exist.

The news was published officially in the summer of 2011: Microsoft was investing a million euros in quantum physics at Delft. And the physics funding organization FOM was putting a million extra into the cooperation between science and industry. Kouwenhoven had already built up a team by then and had started his measurements.

The real news came less than a year later. Kouwenhoven was in Boston in March, where he was addressing the annual meeting of the American Physical Society. I wasn't there, but the news ran through the world of physics like wildfire. Leo Kouwenhoven from Delft in the Netherlands had found the Majorana particle that people had been seeking for so long. Seventy years after it was first theoretically predicted, after a long series of previous attempts to find it. People had looked in some of the strangest places, from the giant particle accelerators at CERN in Geneva to stones from the Antarctic and even the moon. At the presentation, the stairs of the auditorium were full as well, and there were even rows of people standing in the corridors to listen. "Nobel Prize," people were whispering.

Leo himself remained cautious. What he and his team were seeing, he warned, walked like a duck and quacked like a duck. But they still needed more before they could be absolutely sure that it really was a duck. Not that the renowned weekly journal *Science* took much notice of that a couple of weeks later. A dazzling graphic on the cover showed how the quantum wizards from Delft had found the particle: by making it themselves on a nanochip. For this particular occasion, the chip was made of gold and stainless steel. In physical terms, the metallic appearance is utterly meaningless, but it certainly looks pretty.

The picture was made by a physics student from Delft who was also a bit of a graphic artist, and it contains what later turned out to be an unusual inside joke: If you look very closely, on the very tip of a golden nano-bar, you can see the reflection of Ettore Majorana, an image of him that we know from the few photos taken in his short life. A nice homage to one of physics' tragic heroes. And this *was* strictly speaking the news: a glimpse of the Majorana particle had been seen.

When the news came from Boston, my courage as a reporter deserted me for a moment, if I'm being honest. There was something I'd already been afraid of when Leo first whispered the details of his Majorana adventure to me. It was important news, and it had to go in the paper of course.

But how can you explain all that in a couple of paragraphs... Particles that are their own antiparticles, and why that's important for a quantum computer?

What's a quantum computer? I'd been intending to say more about that later, but Einstein and Bohr are looking at me, uncomprehending. So here's the short version: a quantum computer is a calculating device in which information is manipulated in the form of undefined quantum states, instead of being concrete ones and zeroes. This makes simultaneous calculations possible that would otherwise have to be done one after the other. As a result, on paper at least, calculations using this machine are colossally faster.

But back to Delft. To start with, Kouwenhoven's particles aren't real particles in the classical sense, like the ones you can make using a glowing filament or an accelerator. To keep things clear, physicists prefer to refer to them as quasiparticles, collective phenomena in a cloud of electrons, for example, that for the sake of simplicity can be treated as a characteristic of a special particle. It's a bit like a Mexican wave in a football stadium, which isn't really an object, but does have a position and speed.

The researchers from Delft have been able to induce something similar in the special microscopic circuit that they have designed and made in which an indium-antimony nano-wire bridges the gap between two superconducting strips of niobium. The composite whole of the exotic materials, the cooling and a subtle magnetic field mean that the interplay of electrons at the two ends of the nano-wire acquires Majorana-like characteristics. Measurements in the lab in Delft have shown clear indications of this, but not everything is yet certain. Since the discovery in 2012, there is still a question as to whether or not these are indeed the specific Majorana particles that the specialists at Microsoft think they need for a quantum computer. Kouwenhoven is still working on it, now with competing labs from around the world keeping a close and watchful eye on him.

The key property that Microsoft and Kouwenhoven are looking for is what is known as topological protection of the quantum states. This effect would make the quantum bits in a future quantum computer insensitive to external influences. Which is crucial if you're ever going to build a real quantum computer: if they are disturbed even a tiny bit, particles lose their undefined quantum properties, after which you can only do normal calculations with them, at best. Quantum bits – qubits – that can take a knock, that's what they want.

At the time, Kouwenhoven had already tried to explain to me exactly how topological protection works. Not using sums on the blackboard in his office, but with strips of paper and a couple of pencils. "Suppose a Majorana has two energy states," he said. "We'll imagine one of the energy states as a strip of paper with its two ends glued together. You can then twist and deform that ring, but it remains a ring. In the other energy state, the ends have been glued together after a half twist, making a Möbius strip. That's also a ring, but it's topologically different: it only has one side and one edge. The normal loop has two of each, and there's no way of distorting it to make it into a Möbius strip. And that's also the case with the properties of a suitable quantum pair of particles." You can see that mathematically as a Möbius strip, that no amount of twisting and mangling can ever turn into a normal ring. Unless you really use brute force of course, and get the scissors out.

On paper, a pair of Majorana particles is comparable to

the mathematical trick of a Möbius strip, which can never be turned into a normal ring, which in turn is comparable to the state in which the quantum indeterminacy has disappeared. A qubit made of Majorana particles is therefore pretty difficult to disrupt, unless you make a hard and fast measurement of it, for example at the moment when the quantum computer genuinely wants to know the answer to a calculation question.

It's even more difficult to explain precisely why Majorana particles in particular show this kind of almost tangible mathematical idiosyncrasy. But a man's got to do what a man's got to do, of course.

Quantum spin is the crux of the Majorana topology, as seems almost always to be the case in quantum physics. Particles such as electrons have half-integer spins, making them fermions in the jargon of particle physics. They can only form pairs if their spins are opposing. That's the Pauli exclusion principle that we mentioned earlier, which explains the structure of electron shells in atoms. The components of a fermion pair are recognizable, in that one is spin-up and the other is spin-down. You can't confuse the two, and swapping them results in a clearly different pair.

That's different for pairs of bosons, particles with integer spin. They can be swapped over without any difference being visible.

Majoranas are the exception to the rule, however. In fact they are an intermediate form, between bosons and fermions. They are quasi-particles that arise through the collective behavior of electrons, which are fermions that do have their own spin direction and cannot therefore be swapped around *ad infinitum* without any change being visible. Paired Majoranas can therefore only be swapped around in certain ways if nothing is to change. Anything else simply passes them by. That makes quantum pairs of Majoranas exceptionally stable. If it is possible to make them into a qubit, a way of making a stable quantum computer will have been found.

The key thing that Kouwenhoven is looking for in his new experiments is a spin measurement for his home-made quasi-particles. The Majoranas from Delft will only be genuinely usable as intrinsically stable qubits if they have an integer value. Measuring that turned out to be one helluva job, as he tells me by e-mail a couple of times a year whenever I ask how it's going.

Three years later, the lab in Delft has been thoroughly rebuilt to produce a veritable assembly hall for the first real quantum bit, in part thanks to a colossal EU subsidy of fifteen million euros. Experimenters and theoreticians from Delft, Leiden and elsewhere are working on it day and night. In the somber sixties edifice of the Technical Physics department, the QuTech lab is a breath of fresh air, with its light, wood and glass. Refrigeration units are purring softly everywhere and measurements are blinking on screens. The *lingua franca* is English. There are young, ambitious physicists everywhere. Big names from throughout the world are walking in and out.

Leo, who has in the meantime been knighted and whose work is basically seen as a national icon, has not had time for anything else since his Majorana discovery and the EU millions for the first working qubit. Rightly so. Doing research and inventing things where necessary are Leo Kouwenhoven's *raison d'être*. Somebody else will have to write the books.

Einstein nods, understanding. "*Und dass also sind Sie.* That's where you come in."

5
IT'S ALL ABOUT SPIN

IN WHICH ELECTRONS CAN SPIN IN DIFFERENT WAYS AT THE SAME TIME AND HOW THAT CREATES THE POSSIBILITY OF UNIMAGINABLE PROCESSING POWER AND UNBREAKABLE ENCRYPTION.

Niels Bohr definitely does not look amused when he comes back. He and Einstein went to stretch their legs after we'd been talking for hours, taking a tour of the breathtaking Hotel Métropole. Einstein returned a little while ago, undeniably happy, saying that one of the historic meeting rooms in the huge building appears to have been named after him. Not a very big one, on the next floor up, but with parquet flooring, marble, and gold. Enough space for an intimate dinner with maybe sixteen people, or a presentation for a select group. In addition, there is also the Langevin Room, taking around twenty people, and the Marie Curie Room, which can hold about ten.

But that's your lot. There's no Bohr Room and that has

visibly stung its potential namesake somewhat. He fills his pipe crossly.

We really do need to continue and, as a distraction, I use my smartphone to show them a history clip on YouTube, a film that the French physicist Paul Langevin took at the legendary Solvay Conference in 1927 here in this very same hotel. In jerky black and white, we see some of their colleagues: Max Born and Werner Heisenberg, Erwin Schrödinger, Paul Dirac, old Lorentz of course, Bohr himself, and some thirty other key players from the world of physics at that time. They are there at the invitation of the wealthy Belgian industrialist and physics enthusiast Ernest Solvay to talk about quantum theory, whiling away some drab autumnal days in October. Madame Curie comes out of the front door wearing a beautiful shawl, followed by a number of others, including Einstein who has his nose deep in a manuscript written on gray paper. Paul Ehrenfest sticks his tongue out at the camera. Kramers and Bragg at the Grote Markt in Brussels. Prince Louis de Broglie, who is not talking to anybody. Nineteen of the key players will eventually win Nobel Prizes for their contributions, mainly to quantum mechanics. Later, outside on the patio steps, the famous group photo will be taken, with Einstein in the center and Bohr on the far right of the second row.

This photo, I tell them, has always intrigued me because there are so many theorists in it. Which was probably logical at that point in the history of the quantum theory: that strange world at the very smallest levels had just been discovered and absolutely had to be discussed. Yet at the same time, having all those theorists together seemed to make quanta something completely separate from tangible reality. That's perhaps why it's a good idea to clarify once again that quantum theory is simply about physics, about experiments and measurements. With strange outcomes, perhaps, but certainly real enough and you can see them with your own eyes. Quantum strangeness is not something that exists only on paper. It is perfectly real.

And spin is a beautiful example. When Sem Goudsmit and George Uhlenbeck imagined in Leiden in the mid-1920s that an electron could rotate around its axis in precisely two ways, it was a useful insight that solved a lot of theoretical issues. But rotation implied a magnetic effect, so it had to be measurable.

Not at all coincidentally, the Germans Otto Stern and Walter Gerlach had already provided the tools for that back in 1922. It had been suspected for some time that certain atoms have their own intrinsic magnetic field that seemed to be produced by their internal structure somehow. Nobody knew how, back when genuine atomic theory was still being developed. But Stern and Gerlach invented an ingenious experiment to prove it. They fired a beam of silver atoms from a hot oven and passed it between two magnets. And that let them make a remarkable discovery. In the case of classical magnetic particles, the internal fields would have been pointing in all kinds of directions, being deflected to varying extents by the magnetic fields they passed through. On a screen behind the setup, that ought to have produced a smooth distribution of deposited silver atoms, most roughly in the middle and fewer towards the edges. Instead, Stern and Gerlach found two lines of silver atoms, perpendicular to the direction of the magnetic field used. There appeared to be two different types of silver atom, one with the internal magnetic field pointing up and one with it pointing down.

What would later transpire, of course, was that every silver atom has a lone electron in an outer shell and that electron has a spin. That's what the magnetic field in their experimental setup acts on. And electron spin (Einstein and Bohr nod at the same time) comes in two flavors, up and down.

I have to admit that the up and down aspect of spin always feels a bit weird. You instinctively wonder how on Earth an electron knows what is up and what is down. Aren't those concepts a bit too human for a particle that shouldn't really even be called a particle?

That confusion is the core of the quantum mystery, even more so than wave-particle duality, which Richard Feynman in his books called the deepest quantum puzzle. Spin makes the quantum magic literally tangible. Or – for a physicist it's the same thing – measurable at any rate.

In the case of electrons, the measurement is actually done in the same way as in the famous Stern-Gerlach experiment. The electron is shot between the two poles of a vertical magnetic field. If it goes upwards, it is spin-up; if it does downwards, it is spin-down. There are no intermediate variants.

That sounds simple, but the crux of quantum theory is that there is no way whatsoever of predicting what value (up or down) will be measured for any given electron. The result of the measurement is entirely random. The one thing that is definite is that the electron will always have one of the two spin values.

What is also known is the electron in question retains the spin state that was found for it after the measurement; if the measurement is repeated with the same system of magnets, the same result will be obtained as the first time.

But it gets weirder if we go further. Let's have a second measurement that uses magnets that are rotated ninety degrees, so that the field goes from left to right instead of up and down. A measurement then gives spin to the left or spin to the right, again perfectly at random. The measurement of the spin in the up or down direction has apparently had nothing to do with the measurement in the left or right direction.

The real puzzle is when the magnets in the measurement device are then turned again into the original direction and a spin-up or spin-down measurement is done again on the original particle. It turns out no longer necessarily to have the spin that we started with: it is up or down completely at random again. The interim measurement in a different direction has evidently reset the system back to square one.

Physicists know what to do in a case like that: accept the facts. We have to drop the idea of attempting to imagine spin as a genuine rotation of an actual electrically charged sphere, a kind of miniature electromagnet, because it immediately creates confusion. It can't be the truth. The spin, we conclude logically, is an internal property of electrons (for example), an inherent angular momentum that can evidently be expressed in precisely two ways whenever it is measured. The way the spin is measured also inevitably plays a role. More about that later...

Anyone who follows the online lectures by the theoretical physicist Leonard Susskind of Stanford University will see how modern physicists deal with that kind of quantum strangeness: as an inevitable consequence of the type of math that they have cloaked quantum phenomena in. Cause and effect aren't always all that easy to keep apart then, but it's certainly interesting.

A lot of what feels unnatural about the quantum world comes from the fact that there is an essential role in quantum mathematics for the number i, the imaginary number that is the square root of minus one. Those kinds of quantities don't occur in nature, where everything is real. The strangeness arises because what we see in everyday reality is just the projection onto the real world of the complex inner world of atoms and particles. Quite literally. Mathematicians have for a long time referred to numbers in which i plays a natural part as 'complex'. Susskind advises those listening to him not to imagine anything for the phenomena in the quantum world, because our senses are not suited to them. He is probably right, but our sense of reality doesn't always make it any easier.

No matter how strange spin may seem in the real world, it is without doubt the most useful quantum property that there is. The strangeness of electron spin in particular presents a number of unprecedented opportunities in fields such as security, computer capacity, and information storage. Quantum cryptography, quantum computing, and spintronics are the topics we are getting onto, and Einstein and Bohr are getting visibly excited at the prospect. Who would have thought that their theories could have had so much everyday impact?

The field that is furthest advanced is what is known as quantum cryptography, a special form of electronic encryption that is now actually being used in the banking world. And very likely in the military domain as well, but outsiders like us can't get a very good picture of that.

Cryptography is the art of sending messages in a way that nobody else can decipher. Transposition is an example that we all know well from our childhood: shift every letter in the message up in the alphabet by an agreed number of letters and then send the mangled message. The recipient moves it back by the agreed number of letters and the message becomes legible. That works, but it can also be intercepted and cracked easily enough *en route* without detection, simply by trying all twenty-six possible shifts and seeing if a readable message appears.

There are of course smarter types of transposition that make it much more difficult to crack a secret message. These often use a secret key, for instance based on a very big number. The letters in the original message can be replaced by their numerical positions in the alphabet, turning the message into a long series of numbers. Put the long key underneath that series of numbers, several times if necessary, and add all the numbers up vertically to give you a new series that is not at all easy to decipher without the key.

The same applies to the intended recipient, of course, who needs to have the key in order to read the message. The problem of sending a secret message has therefore been shifted, giving you the problem of sending the requisite key safely instead. If an eavesdropper intercepts the key, it can be used easily enough for reading the communication.

All kinds of tricks have been invented over the years, from books full of one-off keys to synchronized encryption devices such as the famous Enigma machine used by the Germans during the Second World War. But all those systems are ultimately still vulnerable to eavesdroppers because the key can be intercepted and the messages can be decrypted unnoticed.

A well-known trick for getting round that is what is known as asymmetric cryptography, which uses two keys: one for encryption and a different one for deciphering. The code for encryption is made public. If the sender wants to send a message, they find the publicly available encryption code of the intended recipient and use it for encoding the message, which is then transmitted. The recipient can then use their own secret second key for deciphering and reading the message. This method can for instance be based on large prime numbers, integers that are only divisible by one and by themselves. It is pretty much impossible in practice to find out what large prime numbers a very big number could be the product of.

Large primes multiplied together therefore make relatively safe public keys. This was used in the 1980s as the principle behind RSA cryptography, which is now used all over the place, for instance in electronic credit card transactions. The system works fine as long as breaking very large numbers back down into their prime factors remains a hopeless task, even for supercomputers. But as soon as the prime factors of a very large number can be found quickly, it is no longer secure and the system is wide open for eavesdroppers and others who are up to no good.

Einstein and Bohr hesitate. What are electronic credit card transactions? And what have quanta got to do with all this?

It's all about being able to take a sneaky look. Because quantum states lose their ambiguity as soon as a measurement is made, intercepting a message in transit can in principle be excluded. Even back in the 1960s, British and American physicists were racking their brains for clever methods of using quantum techniques to send cryptographic keys that would only work if they had not been looked at in transit.

Such methods were not only found, but some banks now

use quantum cryptography as a standard method for securing their data traffic. Customers don't notice anything, but many of them already do their banking using real quanta.

To understand how, we first have to switch to a digital way of thinking. Numbers, including the very large numbers that make up an encryption key, can always be written as a series of ones and zeroes. In jargon, that's called binary: representing a number as a sum of powers of two. A 1 at a certain position in this kind of binary series indicates whether the corresponding nth power of two is present in the whole: in digital terms, 100 means one times 2 to the power 2, plus zero times 2 to the power 1, plus zero times 2 to the power 0, making 4 +0+0, which is four. In the same way, binary 101 is what we call five, 110 is six and 111 is the number seven. It takes a bit of calculation, but a forty-digit number can simply be written in exactly the same way in binary as a series of roughly a hundred and fifty ones and zeroes. If we now call those ones and zeroes 'bits', we've pretty much got the hang of digital jargon. A forty-digit key is about 150 bits in length.

Einstein and Bohr look at me at first as if I've gone nuts. What a palaver just to encode some perfectly ordinary numbers. But then I see the penny drop. Of course! Ones and zeroes... We are starting to get into the discrete domain of quantum, where for instance electron spins can only be up or down when measured. Spins as bits, that's it. How ingenious.

Time to press on. In cryptography, the idea of quantum bits for instance turns out to be a really useful one. To explain it to the gentlemen, it helps to introduce three new cast members, Alice, Bob and Eve. I don't know who invented these characters, but they're definitely useful for keeping things clear when explaining quantum procedures. Eve gets her name because she's eavesdropping of course. But that's by the by.

If Alice wants to send a message to Bob, a key has to be used to make the message illegible for Eve. That key will comprise a series of ones and zeroes, which Alice creates first of all. She does that by using a magnet to make spin measurements on electrons, noting spin up as a 1 and spin down as a 0. After each measurement, she sends electrons on to Bob, whose magnet is oriented in exactly the same way as Alice's and can therefore measure the incoming spins correctly. A series of ones and zeros that Alice had noted as the key will also reach him as long as he has his magnet aligned exactly the same way as Alice.

What about Eve? If she also knows the appropriate orientation of the magnets, she can see the spins passing by just as Bob does and can note down the key of ones and zeroes. Alice and Bob won't even notice that. However, the situation is different if Eve's magnet is not oriented correctly. If her magnetic field is left to right, the measurements she makes will see random up and down spins and produce random ones and zeroes that have nothing to do with the key. If she then passes the spins on to Bob, who is measuring using a vertical field, the series of ones and zeros will again be randomized. If he attempts to use them to decipher Alice's message, he will get garbage. He can quickly conclude that somebody is evidently listening in and Alice and Bob will presumably switch to a different channel.

The weak point in this quantum procedure is the agreement that Bob and Alice make about the fixed direction of their measuring magnets. If that leaks out, Eve can align her magnet the same way and then eavesdrop without the sender or recipient realizing. A bit more ingenuity is needed to avoid that possibility, but it is not impossible to make the communication watertight. To do that, Alice aligns her magnet at random either vertically or horizontally, does a spin measurement, and sends her particle on to Bob. He has aligned his magnet at random either vertically or horizontally as well, and he measures what comes in. They keep on doing that until they've had enough, or at any rate at least until twice the length of the key. Then they tell each other openly what orientation they had used for their magnets for each particle. All the measurements where the orientations were different are then dropped from the series, leaving them with a sequence of ones and zeroes that they agree on. As long as they don't read that sequence out loud, Eve cannot know what it is. The encrypted communication that then follows is secure; if Eve still listens in, Bob will notice it straight away because her measurements will corrupt the message.

The two quantum titans at the table are impressed by the ingenuity of the logic. But Bob and Alice haven't half gone to some lengths to get a message safely from A to B.

I reassure them straight away. In banking practice, for example, the entire procedure is automated and secure communication is possible over tens of kilometers using fiber-optic cables at speeds of more than a million bits per second. Staggering, I admit. But real quanta nonetheless.

Here in the twenty-first century, it seems perfectly natural for information to be stored in the form of ones and zeroes. A computer is little more than a device for reading such digital data, that is then processed by a program to produce new digital data in the form of numbers, words, or images. Our gadgets and devices work so easily simply because of their speed. Modern-day electronics grinds through the requisite processing steps at such a rate that even the most complex jobs hardly take any time at all. We get annoyed if our mobile Internet connection is lost for a moment even if we're right out in the sticks.

But it can go differently. Some tasks are fundamentally out of reach for even the biggest supercomputers. We're not particularly talking about arithmetic. Any computer can do sums and even people can sometimes do pretty well in their head or on paper. The problems that are really tricky are the ones that have a lot of potential solutions and each and every one of them has to be checked to see if it might work. Searching is something computers aren't much good at all. Factorizing big numbers is a prime example. Mathematicians have been arguing for years about whether extracting prime factors from their product merely gets exponentially more awkward as the product gets longer, or whether it's even worse than that. But in both cases, you can safely say that cracking a secret code with prime factors is obviously pointless if the calculations would take longer than the current age of the universe.

It is precisely the built-in vagueness and ambiguity of the quantum world that, on paper at any rate, could make calculation using quanta such an unbelievably powerful tool. Instead of tackling a problem case by case, a quantum computer could in theory explore and calculate all possibilities at once, filtering the solution out from them.

That vagueness is already inherent at the very smallest level, in the electron spin (which incidentally has the value one half), for which the direction (up or down) is only determined when the measurement is made. On the other hand, that is also confusing. How can information be represented or processed using units that don't even know if they are a zero or one?

In principle, that doesn't have to be a problem. In a digital computer, information is currently stored digitally in the states of transistors, but that could equally well be electron spins, for instance. What classical computing calls 'bits' are known as 'qubits' in the quantum case. The spin of an electron is indeterminate as long as it isn't measured and the state is vague. But if a measurement is done on it, for instance to see whether the spin is up or down, that property is then fixed as whichever value was measured. If the observed spin was up, it can in principle be flipped using a magnet, changing our imaginary qubit from a zero to a one. In an ordinary computer, sequences of bits are processed using additions and comparisons, changing some bits from ones into zeroes and vice versa. The technology will undoubtedly be much more cumbersome in a computer using spins and magnets than in an electronic one full of chips, but in principle the same sums can be done. The question is of course why on Earth you would want to bother, if ordinary computers can already do it.

The strength of calculation using spins lies in the indeterminacy that is present until the measurement is made. A spinning qubit that is left undisturbed exists in both the up and down states at the same time, and is therefore simultaneously both a 1 and a 0. A logic process that turns every 1 into a 0 (known in the jargon as a NOT operator) in a quantum computer would literally mean turning the magnetic field in the measuring system around. Cumbersome maybe, but it's possible, and it actually does two calculations in a single step. Something comparable is also possible for the other logical operators, the building blocks of every computer program. What it comes down to is that a quantum computer carries out a complicated series of inversions in the measurement apparatus for various spinning qubits, but without any observations actually being made. The program then keeps manipulating the states of the spins that the data has been encoded into, all at the same time. The trick of quantum programming is to end up so that a measurement of the system will give the required answer. That is a problem in itself, because a quantum observation only turns one of all the possibilities in the superposed states into reality.

In the meantime, rather classical-looking algorithms needed for quantum calculations have been developed by Peter Shor and David Deutsch, among others. Experiments have also been carried out that genuinely demonstrate a quantum calculation, including one using seven qubits at once to determine the prime factors of the number fifteen. Okay, we can all work out in our heads that the answer is three and five.

So it's only fair to concede that the great breakthrough of quantum computing is yet to come. There are plenty of ideas and plenty of proposals for ingenious quantum systems that could provide the qubits needed. Hundreds of millions are being poured into research worldwide, and Delft University of Technology is in fact seen as one of the centers of expertise in the field. Together with the University of Leiden and with the help of EU mega-subsidies, it is putting a lot of effort into producing reliable and practical qubits. An important point is not only that this system has to be sturdy, but also that any errors have to be spotted and corrected in time. A quantum computer that only works a little is less use than a digital computer that works properly.

At the same time, people are keeping a very close eye on the American-Canadian venture D-Wave in Ottawa, which claims already to be able to deliver a working quantum computer. Microsoft, NASA, and others have paid big money for a device that according to the specifications has 512 superconducting qubits, so that they can then put it through its paces.

As yet, the international research community is unconvinced. A Swiss test project by IBM concluded that if there are any quantum-like calculations in the D-Wave computer, they are no quicker than in a normal electronic computer – slower in fact. D-Wave, which employs a number of acknowledged experts in the field, called the tests misleading and insists that it *can* deliver a quantum computer.

The biggest misunderstanding about quantum computing however, is that it can be compared to the work done by ordinary computers. All the experts emphasize that quantum calculation will always depend on difficult and complex procedures. Every calculation is essentially a physical experiment, needing a lot of preparation and careful work. During the time that a quantum computer takes a single step, an ordinary supercomputer will have been able to solve most calculation sums.

That does not apply, though, for needle-in-a-haystack problems. A phone number can be found in the directory simply by reading through it lightning fast from cover to cover, the best way through a maze by trying all combinations of paths one by one, a favorable protein structure by examining every possible conformation. As the problem becomes more complex, the calculation time required goes up disproportionately. That's where the quantum computer comes in: It is in principle capable of looking at all options in a single calculation step and finding the optimum. That one calculation step can take weeks or months of preparation if necessary, but ultimately the answer is obtained much more quickly than by the brute force approach.

In Delft and Leiden, they suspect that quantum computing will become something similar to conventional supercomputing now: a service that university computer centers will offer for large calculations, particularly for specialist users from science and industry. They will be used for modeling, picking through enormous datasets, and doing simulations – even simulations of quantum processes, they think. A word processor or smartphone using qubits is not only almost certainly practically impossible, but it is utterly pointless even to want one.

Not for the calculation power, at any rate. Pure quantum technology does however seem likely to end up on our desks and in our pockets by a totally different route. Over recent years, the field known as spintronics has been making huge advances. And it is possible (I look at my table companions for a moment) that diamonds will play a crucial role.

I can't resist it. I asked Einstein directly whether he saw Marilyn Monroe in 1953 in the otherwise pretty abominable film *Gentlemen Prefer Blondes*, in which she sings 'Diamonds Are A Girl's Best Friend'? I also tell him that I associate him with another film from long after his death, Nicolas Roeg's 1985 work *Insignificance*. The scene where Theresa Russell as a voluptuous Monroe in her famous white dress is explaining relativity theory to an embarrassed Einstein (Michael Emil) using trains, torches, and a pocket watch can still be found on YouTube. Her summary at the end is that all measurements of time and space depend on the specific observer.

Einstein reaches for my iPhone, seemingly having worked out how to use the device. "Where is the YouTube app?" In the meantime, I tell Bohr the story of spintronics, insofar as there actually is one. There are plenty of ideas. Virtually all the electronics around us are made of silicon, a wonderful semiconductor that can be found in large quantities on Earth. Silicon as a material is superbly suited for directing electrons around the place. We can store and transmit signals using tiny currents and charges. If there is an electron just there, that bit is a one. If not, it's a zero. Directing those electrons from place to place costs energy; definitely one disadvantage of the technology. But there's another much more essential problem with silicon: we're approaching the point where it will be impossible to make electronic circuits any smaller. Once individual silicon atoms become involved, the semiconductor nature will cease. And if the circuits can't be made smaller, the electronics can't be made quicker.

But why would we necessarily want to base the information processing in computers and gadgets on electrons whizzing about? The last ten years have seen an alternative emerge, after an unusual property of diamonds was discovered. It is possible to introduce crystal errors at the atomic scale in the fantastic and super-strong crystal structure of carbon. This is done by firing nitrogen atoms into the lattice. They knock a carbon atom out of position and end up embedded just next to the open spot. This creates a kind of trap in the lattice for individual electrons, not really free electrons, but free enough to be able to do their own thing in quantum mechanical terms. The spin of these electrons, for instance, turns out to be easy to control and query using light or magnetism. That makes the spin available for use as a qubit at room temperature. The price of diamonds is not even a significant restriction. Diamonds used to have to be dug out of dangerous mines, but nowadays diamond can be grown for a few dollars a carat in the lab from methane – bog-standard natural gas.

Spintronics is still in its infancy and many of the ins and outs are not yet very clear even to enthusiastic researchers. But there is a significant chance that somewhere in the future, our smartphones and the tablets in our bags will not use electrons whizzing about but will instead be based on quantum spins. So maybe it's better to say that diamonds may be everyone's best friend.

"Amazing," says Einstein, in exactly the same way as at the end of the Einstein-Monroe scene in *Insignificance*. Bohr nods. The reality of spin is amazing in all kinds of ways.

6

AN ORAL EXAM WITH GHOSTS

HOW QUANTUM PARTICLES TURN OUT TO STILL BE ABLE TO SENSE EACH OTHER INSTANTLY, SOMETIMES AT INFINITE DISTANCES, AND WHY THIS DOES NOT TURN OUR WORLD OF CAUSE AND EFFECT UPSIDE DOWN.

It's getting later without us even noticing. The other guests in the dimly lit Brussels brasserie Métropole are now drinking large glasses of beer or cognac, more and more often without the coffee. Every now and then, the softly swaying crystal chandeliers are casting glints on the decorative wallpaper and marbled table tops. People are laughing. I'm even in the mood for a cigar, but I restrain myself.

"We should really talk about 1935 now," I say. I mean the famous EPR paper, which seemed to have put a bomb under the entirety of quantum theory for a little while. It was named after its authors, Albert Einstein and his two assistants Boris Podolski and Nathan Rosen, three physicists from Princeton who made a major intellectual effort to show that something was fundamentally wrong with the by then more or less generally accepted ideas of Bohr and his Copenhagen school about the reality of quanta. And the argument was a favorite ploy of great physicists: it took the form of a paradox, an impossible conclusion that naturally destroys the principles underpinning a line of reasoning.

In this case, we had to choose which we would prefer to believe: that the quantum world could circumvent space and time, or that any theory suggesting such a thing needed to be worked out in much more detail.

Einstein is grinning, Bohr is sitting up straight. *Natürlich*. Of course, the EPR paradox. Here, in the Hotel Métropole, they fought endless intellectual battles during the twenties about the true nature of quantum mechanics. Einstein attacked: God doesn't play dice. Bohr countered: God's got nothing to do with it. He seemed to have won on points, but Einstein attacked again, using other arguments. It was maddening sometimes, even for Bohr, who himself was particularly known for his ability to exhaust his opponents with his arguments. But the endless quantum debate sharpened the thinking on both sides. And the EPR paradox, which was published in 1935, was the crowning glory. One of the strongest Einsteinian thought experiments ever, the ultimate make-or-break test for Einstein and for the quantum theoreticians.

At this point I'm starting to get the unpleasant feeling that my monolog on the reality of the quantum, which was intended to end with the conclusion that quantum mechanics is actually not that difficult to understand at all, is degenerating from a narrative into an oral exam. After all, the men in front of me are the founders of what I am going to discuss now, the creators of the arguments, the ifs and buts, all the details that you can think of about quantum reality. What is there to explain to them?

I order three cigars and an ashtray and decide to go for it: my demonstration of higher quantum acrobatics. I hope I won't be making an idiot of my humble scientific journalistic self during it. Fortunately, there's a safety net up ahead, created by other big names like John Bell, Alain Aspect, and Leonard Susskind. They will turn out to be great company.

We light our cigars.

The paradox that was formulated by Einstein and his associates in 1935 is based on a thought experiment involving two quantum particles. It was already indisputable by then that such particles exist. According to quantum mechanics, a particle such as an electron with a spin is in two states at the same time, as long as it is left well alone. When the spin is measured, though, one of the two possible spin values becomes the reality. It's rather like a dice that's still in your hand: it is capable of giving all values from one to six spots, but will only give one result after it has been thrown. The quantum theory of that moment was purely about the results of throwing dice; how dice actually work does not matter. Einstein refused to accept this, wanting to know why the imaginary dice falls the way it does, on the assumption that it simply involves the laws of cause and effect. He believed that reality could not be a matter of coincidence.

To demonstrate this, he chose a tried and tested method, the concept of a paradox. He imagined two particles that are quantum-mechanically entangled, i.e. they jointly acquire some quantum property or other. In theory, all possible results of a measurement on it are therefore available at the same time. Eventually, the measurement will select one of these results.

But this measurement is carried out in a special way. The two entangled particles are separated physically and set up for measurement far from each other. So what happens when the states of the particles are actually measured?

The classical reasoning is simple. Suppose that a pair of electrons, which each have a spin of a half, have been created from a particle with spin zero. The total spin is zero and remains zero. So if measurement shows an up-spin for one particle, a measurement on the other particle must give a down-spin in order to conserve the net zero result. In practice this doesn't seem to be a problem, but quantum mechanics does have a lot to explain. It describes both electrons as spin-half particles that are both in every conceivable spin state until a measurement is carried out. Einstein's threesome wondered how a measurement of one particle, which therefore selects one specific spin state, could set the state of the other particle even before a measurement has been carried out on it.

The trio concluded that information is evidently passed from one measurement to the other. But if that's true, there was another problem: information can travel at the speed of light at the most. So if measurements are carried out on both electrons at exactly the same time, it is impossible for either electron to know what value was measured for the other one. Yet quantum theory predicts that the results are linked together. In other words, in this thought experiment, quantum theory seemed to be totally ignoring one of the basic laws of the universe, as described in the theory of relativity (which just happened to be Einstein's). There seemed to be no alternative.

If ever Bohr had been succinct about anything, it had been about this. "It's not a deathblow," he answered after thinking for a long time. "In the final analysis, it's Einstein's problem if his pet theory of relativity gets into trouble when the EPR experiment gives the expected result." The problem that two particles could be almost infinitely far apart from each other yet be aware of each other's state was not an issue, because no actual information could be sent at all in the experiment as described. The observer of the one electron could not make anything clear to the observer of the other one. The first measurement would enforce the random selection of one of the possible states. The second measurement elsewhere can confirm it if you want, but doesn't say anything more. Even if the observers had agreed that the first of them would return home immediately the moment the first spin-up measurement was obtained, the other could only assume that this was really happening. If the person who traveled into deep space didn't stick to the agreement, the one who stayed at home could only realize that much later.

Tight-lipped faces and mumbling at the table. Einstein and Bohr seem fully prepared to start launching intellectual attacks on each other again. Thank goodness, I still have an ace up my sleeve, the Frenchman who actually carried out that thought experiment in the eighties. It does not matter much that both protagonists were already dead by then.

Alain Aspect is without any doubt the physicist with the most impressively twirling mustache, but that is beside the point. He is also an excellent experimental physicist. In his laboratory in Paris, Aspect let pairs of polarized photons travel some distance away from their common source, after which he measured the direction they were polarized in. The measurements always matched each other perfectly: if one was up, the other was down. No way round it. Einstein's spooky action at a distance really existed. A measurement of one particle never conflicted with a measurement of the other.

So Einstein was therefore wrong. For sure, the connection between two entangled quantum particles is most remarkable. It's as if you're asking one of a pair of twins random questions and writing down the answers, and then interview the other twin too, only to discover that they always give opposite answers to all questions. Non-physicists are quickly inclined to believe that it's all telepathy. But it apparently is the way that God has set things up: he does play dice, to quote that clichéd phrase once again.

But we'll get back to that later.

First let's meet Leonard Susskind, the man who in 1970 was the first to think about what would happen if you picture quantum particles as vibrating elastic bands, making him the founder of the famous string theory.

One of the nicest moments in the challenging series of online lectures about quantum mechanics that this American theoretician has given at Stanford over recent years was his argument about precisely that experiment by Einstein, Podolski and Rosen. While taking a bite of the cake that was in a bag on the lectern in front of him, as always during his lectures, he explained that EPR actually has nothing to do with quantum theory. Let alone with the foundations of physics. The paradox, said Susskind, munching slowly, also exists in a fully classical system.

Take a hat, he suggested, and put two different coins in it. A nickel and a dime (or for the Europeans, a five euro-cent piece and a ten euro-cent piece). Ask two students – let's call them Alice and Bob for the sake of convenience – to take out a coin without looking. After that, we send Alice to the Andromeda Nebula or beyond, while Bob stays at home in California. He surfs a bit, goes to college, graduates. With the coin in his pocket, unseen.

Meanwhile, Alice makes a 'measurement': she finally opens her hand after years of traveling among the distant stars, and sees what she had not known all that time: she has the five-cent coin. She instantly also knows what Bob has got, back on Earth: the ten-cent coin. Alice's measurement not only tells her which she has got, but also seems to determine what Bob will see if he makes a 'measurement'.

Is that a violation of the theory of relativity? No, of course not! Although Alice now knows what Bob has got, no information has been transmitted in any way. Alice can send Bob a message to tell him what he has got in his pocket, and receive a confirmation from him. But the information just travels at the speed of light, at the most.

This is how we should consider Einstein's paradox with the two particles, said Susskind in his YouTube lecture. Relativity is not the odd thing about the Einstein-Podolski-Rosen paradox. Relativity isn't violated anywhere. The odd thing, if there is one, is in the quantum measurements. The entangled particles are both in an ambiguous summation of possible states. If one of them is measured, a state is selected for it. Yet at the same time, a state is selected for the other one too, because they are linked to each other according to the immutable law of quantum entanglement, like heads or tails on a five-cent coin. In the mathematical equations, which describe it better that any intuition, the distance between heads and tails does not play a part. It is purely about the mathematical but indisputable link between the two.

In his lectures, Susskind patiently fills one blackboard after another on the subject, while emphasizing that the trouble we have getting a feeling for quantum phenomena is caused by the weird math of quantum mechanics. This includes not only normal numbers, but also imaginary components, in which the square root of minus one has a crucial mathematical role. The message is that imaginary components of the equations do not yield any measurable quantities, but behind the screens they certainly do affect the results of measurements. Physicists have got used to outsiders who keep thinking on classical lines being surprised by it. Quantum behavior is a bit like the tricks that a conjuror does: incomprehensible miracles are taking place, but in reality they rely on sleight of hand and hidden pockets and strings.

But if the Einstein-Podolski-Rosen paradox actually applies to common or garden coins, does this mean that everything is fine for pairs of quantum particles? In the nineteen sixties, the Irish theoretical physicist John Bell was convinced that quantum theory was not the whole story. Was there perhaps a deeper reality, underpinned by classical variables that are not directly visible for quantum theory? If there were hidden variables, quantum vagueness would become something like the temperature and pressure of a gas, an effect resulting from predictable processes (colliding molecules, in the case of the gas), of which there are simply too many to follow.

It turned out that the idea of the two entangled particles could be a handy model here too. Bell designed an actual test that could determine whether the states of paired quantum particles are always randomly assigned. If not, there would evidently have to be a deeper reality that we may not have access to, but which is dominated by more classical laws of cause and effect.

I realize that the Bell test is very difficult to explain in popular science. So much is combined all at once that it is hard to think of a good starting point. But the two men at the table can hardly wait. They were already dead in the sixties, so it's really a great pity that they never met Bell. So bring it on.

Again, I get this feeling of being in an exam.

I start. "Suppose you have a hydrogen atom that is pulled apart into two identical parts..." One half zings off to the left into a setup in the lab, the other half whizzes off to the right. The initial atom did not have any spin. But the fragments, whatever they may be, can have spin. Their spin is what it is all about. If the total spin before the fission was zero, the total spin afterwards is zero too. But as always in the quantum world, which half has what spin is random. The amazing thing about the example is that, although the measurements of the particles do yield a random result, they must add up to zero together. If one particle is measured, there's not much that's random about the value for the other. No wiggle room at all, to be precise.

A classical physicist can buy that. It could be because the spins are determined from the moment of fission, but in some hidden way. Imagine a watch face attached to each particle, for instance, where the hands are put in whatever position during the fission, but always in such a way that the two hands point in opposite directions. If it is a quarter to the hour on one particle, then it is quarter past on the other one (the hours don't matter here).

What matters now are the measurements. A device is used for this that compares spins against its own vertical axis. If this device sees a dial pass by with the hand in the upper half of the watch face, it registers an up-spin. In the other case, it registers a down-spin. If a second detector further along looks at the dial of the other particle, the secret hand there will be pointing in the diametrically opposite direction. If this detector is using the same vertical references as the first, the measurement will therefore be opposite. If the left-hand particle is up, the right-hand particle will be down. And vice versa.

So far, everything seems to be okay. This is what we're used to: objects have fixed properties.

In 1984, John Bell, who at that time worked for CERN, the laboratory for particle physics in Geneva, asked a question that was to change everything. What would happen if the second detector was not in a neatly upright position? What if it was tilted? At first sight, this seems a silly idea. But the results turn out to be so earth-shattering that it's well worth accepting the lack of logic.

If the second detector is tilted, the observation of the passing particle's secret dial changes. The upper and lower halves are no longer what they were.

According to quantum theory, this doesn't matter at all. If one measurement shows spin-up, the other should show a spin-down. But this is different for the hands on the watch face. Bell showed that the same results as predicted by quantum theory would be found in a few positions, but things would be completely different in all other cases. Bell formulated that difference as an inequality that would become famous later on. The inequality always holds true in a classical world. In a quantum world, there are exceptions to it.

"So then?" Einstein looks at me inquisitively over his cigar. "What's the conclusion? Who was right?"

I hesitate. Who would want to contradict the greatest

physicist of the twentieth century? But a man's got to do what a man's got to do. "I'm afraid, Professor Einstein, that you were wrong." John Bell was able to show that no theory whatsoever could give the quantum-based results by invoking dial-like, hidden, local markers. Local hidden variables simply cannot account for the quantum world. More than that, even: Bell also showed that hidden variables may play a role, but that they would have to do so across the whole system, even if its ends are light-years apart.

Bohr cannot help grinning. He may not be quite such a great physicist as his good friend Albert, but at least he is right on this one. And that helps. And that was the reason why he finally won, over the course of two decades, every debate with his stubborn colleague. He puts a big warm hand on Einstein's shoulder.

The scene only lasts a few seconds. Then Einstein is ready to ask the inevitable question. Straightening up again, he asks, "What does Nature have to say about this? After all, everything up to now was just a *Gedankenexperiment*. Someone has to perform the experiments." Someone has to determine experimentally which of the two it is: do we live in a classical world or in a quantum world?

For this, we have to go to our friend Alain Aspect. He was the first to take Bell's considerations and use them literally as a recipe for an experiment, actually testing the famous inequality. If he could prove in the reality of the lab that Bell's inequality is violated, it would determine in one fell swoop that all the ideas about hidden variables and classical shadow realities belong to the realm of fairy tales.

In 1981, after carrying out endless preparatory work in Paris, he performed experiments with paired photons from a calcium source and reached a clear conclusion. Bell's inequality would demand a result of less than 2 for his tests. It did not apply: a value of 2.697 was obtained. The test was a success and therefore quantum mechanics held water. We live in a universe dominated by the laws and logic of the quantum.

The old way of thinking is finished and the old ghosts have been banished. At the time, not a single newspaper wrote about it. Too complicated. It didn't affect anyone.

At long last, the case of the quantum ghosts seems to have been settled. At the same time, it remains counterintuitive and odd that physicists keep wanting to see it with their own eyes. Numerous experiments have been done since that time to get a better understanding of the subtleties of remote quantum entanglement, always leading to the same, unnatural result: the bottom line is that the ghosts exist. The largest version to date bridged 144 kilometers between La Palma and Tenerife. But the devil is in the details, and one of the physicists who wanted to be sure about it, was the Delft researcher Ronald Hanson. He was educated in Groningen and worked in America for some time.

Over the past few years he had been involved, as a professor, in an experiment with entangled particles that would overshadow all previous experiments, in size if nothing else. Hanson used the entire campus of Delft Technical University as a test setup. Quite literally: its ends are approximately a mile apart, as the crow flies.

One end was in Hanson's laboratory, in a cellar underneath the Technical Physics building, where laser light flashed through labyrinths of lenses, mirrors and prisms, finally exciting tiny little diamonds into the appropriate quantum state. The other end was two kilometers away, on the other side of the highway, at the Delft Reactor Institute

(DRI), where there is a similar setup. The two ends were connected by fiber-optic cables underneath Mekelweg, the central green avenue for trams and cyclists on the campus.

The first objective of the experiments in Delft was what is known as a loophole-free Bell test, in which the remote entanglement is determined for all pairs of particles that are created. The initial experiments of Aspect, for instance, left the possibility open that his results had been a statistical fluke. Hanson was going for one hundred percent reliability, although not merely to show once again that Bell was right. That had already been done frequently since 2001. Hanson had bigger fish to fry: teleportation.

What Hanson's tests are all about actually is passing quantum states on from one part of a system to another, without the state being broken. Kilometers do not have to be bridged for these tests, but bridging meters is important, particularly for the largest project that Delft is carrying out in quantum research: building the first quantum computer. Tens of millions of euros, some of it awarded by the EU, have been used to set up an entire wing of Technical Physics as a quantum lab.

Postdocs and PhD students from various Dutch universities are sitting at desks there, while their cryogenically cooled experiments take place one floor below. They sometimes write very photogenic complicated equations on the glass partitions between the departments. This is all about building a quantum computer. The first objective is what is known as a reliable qubit. Where digital computers use zeroes and ones – bits – for their calculations, quantum computers use quantum bits, or qubits for short, which are a superposition of both one and zero as long as nobody looks at them. This sounds like a recipe for being incredibly vague. But physicists have known for decades (particularly because of the work of IBM mathematician Charles Bennett, someone who is tipped for a Nobel prize almost every year) how smart manipulation of quantum bits can result in unprecedented computing capacity. In theory, a quantum computer is almost infinitely faster than a digital one.

In newspapers and on television, quantum computing is always talked about in the same breath as cracking cryptographic codes and other calculations for which even today's supercomputers are too slow. However, physicists like Hanson and his Delft colleagues Leo Kouwenhoven and Lieven Vandersypen can hardly wait until their machine brings quantum behavior within easy reach. Much of the quantum behavior that is still theoretical now can be simulated once that happens. And that is just in the lab in Delft.

We have discussed quantum computers earlier. Where Hanson's entanglement work comes in is the point at which the various qubits in a quantum computer have to cooperate. Quantum technology is in fact the only solution for this. After all, when measurements are made on quantum systems, the quantum magic – its inherent vagueness – will be disturbed, after which nothing but a conventional calculator remains. Components of a quantum computer must be linked to each other in ways that leave the quantum world intact. Teleportation seems like a solution, and it may even be the only solution. Quantum teleportation, to be precise.

This has been possible since the seventies in science fiction. Captain Kirk in the famous *Star Trek* series only has to use his advanced communication watch to call his flight engineer Scotty, and he is transmitted back to his starship immediately, usually just in time to escape his alien attackers. We see how the chubby space captain in his tight-fitting uniform fades away on one location and then rematerializes back in the Enterprise. A neat trick. Not a single atom of Kirk has disappeared or been lost *en route*. Nothing special for the captain, though.

"Beam me up, Scotty!" has become a common phrase in popular science stories, and in stories about quantum physics too. But in this case, Scotty is the Dutch professor Ronald Hanson, and Captain Kirk is not a human being but instead usually a slightly excited lattice state in a diamond.

In 2014, Hanson's team already showed, on an optical bench in the cellars of the lab on Mekelweg in Delft, that quantum teleportation does exist. *Science* published their work, a sign that something essentially new had been achieved. To start with, the Delft physicists created pairs of entangled light particles for which the overall joint state is known, but not the states of the two constituent parts. Those photons are separated from each other by a distance of up to a few meters and stored in special diamond lattices. Up to this point, the experiment is the classical EPR entanglement experiment, which we already know to be possible: two particles that remain linked to each other quantum mechanically so that making a measurement on one determines the state of the other as well. That's been shown dozens of times and so it's not new, amazing as it may be to see that it really works like that.

However, Hanson wanted to use this remote linkage to transfer quantum states from one location to another, undamaged. To achieve this, an atom is brought into an excited state in one of the diamond lattices. The stored photon senses the change and is quantum mechanically linked to the excited atom: together they form a new physical system. In the other tiny diamond, the entangled photon senses the changes that its partner is undergoing elsewhere, and it also becomes linked to an atom. The state of that atom in the second diamond lattice is therefore related to the state of the excited atom in the first diamond; the information from one atom has been transferred to an atom at the other end of the optical bench via the quantum channel.

The reason why the leading journal Science particularly wanted to publish the work done at Delft is that Hanson's system did not give a mere statistical probability. It always works. If anything goes wrong in the entanglement, the transfer will be interrupted and the system will keep on trying until it does work. Previous quantum teleportation experiments, some of which attracted the attention of the media with claims about teleportation using satellite channels over hundreds of kilometers, are not foolproof. Hanson has maintained for years in every comment on such claims that this is a fatal flaw if you want to achieve a practical application. He is the only one who knows how to teleport without any mistakes. As the readers of *Science* have also known since 2014.

What Hanson also knows is how recalcitrant such techniques can be in practice. Since 2014, the Delft team has been working enthusiastically on setting up the immense experiment that runs from one corner of the campus to the other. Laser setups have been constructed, measuring equipment has been developed, kilometers of glass fiber pulled through tunnels that most TU staff do not even know exist, let alone how to find their way through them. When exploring the options, Hanson was mightily relieved to meet the one and only maintenance technician who knows his way around them, literally just sticking his head up out of a manhole. But the measurements are on the very limit of what is achievable, and have already taken years longer than was initially hoped.

The distance of kilometers in the Delft experiment is essential, because they need to show that transferring the state of the atom from one diamond to the other takes literally no time. In theory, teleportation always has a classical loophole, which makes it possible for one system to affect the other by signals at the speed of light. The transfer in the quantum variant is instantaneous: when one changes, so does the other. No matter what the distance. The distance in kilometers between the entangled parts of the system has to be sufficient for any difference in time between the changes at the two ends to be measurable. Accurate clocks and fiber-optic cables are needed for proving this.

We stop talking for a little while.

I take a special edition of *de Volkskrant* supplement called *Sir Edmund* from my bag. It is dated Saturday, October 24, 2015. The cover prominently shows a black-and-white drawing of Albert Einstein's head. The caption underneath says "The Spirit of Delft." The title refers to a six-page comic strip by underground illustrator Erik Kriek. It's about the quantum experiment in Delft performed by Ronald Hanson and his smart, laconic PhD student Bas Hensen.

Einstein opens the supplement suspiciously. "Uh-oh," he seems to be thinking. "Have the newspapers become picture books today too? Even when they're about science? And why am I on the cover and not the young *Herr Professor* from Delft?"

In the summer of 2015, I picked up the first signals from Ronald Hanson that big news was about to come out about his quantum experiments on Mekelweg. He was going to submit an article about the results to the British journal

Nature and if it was published, the news could be published in our newspaper too. We decided on an unusual format for reporting: a comic strip about the first loophole-free Bell test. How a group of Dutch physicists removed the remaining doubts about the remarkable quantum universe that we live in. After all, it was kind of like a boy's adventure story.

Early in the summer of 2015, Hanson and Hensen stayed in the lab for two weeks and collected a total of 220 hours of measurements. They continually put their diamond lattice in one lab into one quantum state, and linked that state to the diamond at the reactor institute a mile or so away, while trying to measure the spin state in a random direction over and over again and comparing the measurements via fiber-optic cables.

The trick was mainly to filter the measurement statistics to leave just the very few occasions that everything ran smoothly. Such subtle measurements were only successful approximately once every hundred million times. But that's sufficient. Weeks of work yielded 245 successful measurements that taken together substantially exceeded Bell's famous threshold value of 2. In *Nature*, Hanson gave a value of 2.42 plus or minus 0.20. The random measurements at the two ends of the setup were more closely interrelated than classical physics was able to explain.

During the days after the publication, Ronald Hanson seemed to be everywhere in the media, who were mainly keen on being able to say that even Albert Einstein can be wrong. Most of the viewers and readers missed the nifty physics of the Bell experiments in Delft. And the odd physicist was still grumbling that 245 measurements did not seem very much. If, against all expectations, Hanson were to have found that the ghost link did not exist, he would probably not have got away with it, the critics reckon. But he has now.

We put our cigars out at almost the same time. Three stubs smoldering in an Stella Artois ashtray.

Einstein is grumbling. Engineers who are tinkering with entanglement and using quantum bits in their calculations are all very well, but the essential question of whether quantum theory tells the whole story has not been answered, of course. Always trust your intuition, he says, tapping the table top with his index finger. When Eddington used photos of the solar eclipse in 1919 to show that his theory of general relativity was correct, someone had also asked him, "And what if it hadn't been proved?" He could only think of one answer. "Well, that would be a pity for God, but my theory is still right."

Bohr looks up demonstratively. "Albert, please stop it, *mein Lieber*. If they're now managing to use quanta in their calculations, why are you still whining about theory?"

IT'S THE LATE SUMMER OF 2015 AND THE PHYSICIST ***RONALD HANSON*** IS SITTING AT HIS DESK IN DELFT. THE NATIONAL AND FOREIGN MEDIA WANT TO KNOW ALL ABOUT IT: HOW HAS A YOUNG RESEARCHER FROM DELFT MANAGED TO PROVE DEFINITIVELY THAT ***ALBERT EINSTEIN*** GOT IT ALL WRONG?

THE SPIRIT OF DELFT

TEXT: MARTIJN VAN CALMTHOUT · DRAWINGS: ERIK KRIEK

AFTER A NERVE-WRACKING SUMMER IN A DARKENED BASEMENT FULL OF LASERS AND MIRRORS AND WEEK OF UNCERTAINTY ABOUT THE SCIENTIFIC ARTICLE THAT HE SUBMITTED TO THE PRESTIGIOUS MAGAZINE ***NATURE***, THERE'S FINALLY SOME GOOD NEWS: THE ARTICLE HAS BEEN ACCEPTED BY SOME HEAVYWEIGHT COLLEAGUES. THE PROFESSIONAL MEDIA HAVE GOT WIND OF IT AND THEY CAN SNIFF A SCOOP.

STUDENTS ARE CYCLING AROUND THE DELFT CAMPUS OUTSIDE, WITH NO IDEA THAT THERE ARE FIBER-OPTIC CABLES UNDER THEIR FEET - CABLES THAT ***HANSON*** USED FOR A TOUR DE FORCE THAT IS GOING TO BE SCIENTIFIC WORLD NEWS.

A COUPLE OF MINUTES' BIKE RIDE FURTHER ALONG IS THE TECHNICAL UNIVERSITY'S ***NUCLEAR REACTOR***. BEHIND A HEAVY STEEL DOOR IN THE BASEMENT IS ONE HALF OF A GIGANTIC MEASURING DEVICE THAT HANSON BUILT WITH JUST ONE THING IN MIND...

... TO FIND THE GHOST THAT ***EINSTEIN*** FOUND SO ABHORRENT.

THE 1930S WERE WILD YEARS FOR PHYSICS, WHEN YOUNG GENIUSES LIKE ***WERNER HEISENBERG*** WERE DEVELOPING QUANTUM THEORY.

IT DESCRIBES THE WORLD OF PARTICLES. BUT THAT TURNS OUT TO BE A VERY STRANGE WORLD. IF THE THEORY IS RIGHT, EVERYTHING IS ***VAGUE*** AND ***UNDETERMINED***. UNTIL IT IS MEASURED. IT ONLY GETS A REAL VALUE THEN – AND WHICH POSSIBLE VALUE IT GETS SEEMS TO BE RANDOM.

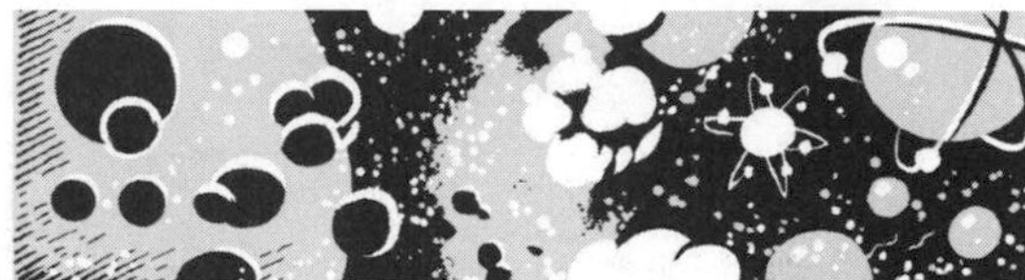

THAT'S A REAL MYSTERY. A SPINNING TOP ROTATES ONE WAY OR THE OTHER – FAIR ENOUGH. BUT AN ***ATOM*** OR ***PARTICLE*** THAT IS ***SPINNING*** SEEMS TO BE GOING BOTH WAYS! A MEASUREMENT SHOWS ONE VALUE OR THE OTHER, BUT WHICH ONE SEEMS TO BE A MATTER OF PURE CHANCE.

EINSTEIN DOESN'T LIKE THAT AT ALL. HE DECIDES TO DEMONSTRATE WITH A ***PARADOX*** ONCE AND FOR ALL THAT QUANTUM THEORY IS A FAILURE.

HE IMAGINES TWO PARTICLES THAT ARE KNOWN TO BE SPINNING IN ***OPPOSITE*** DIRECTIONS, BUT NO MORE THAN THAT. ACCORDING TO ***QUANTUM THEORY***, THE SPIN OF EACH PARTICLE IS NOT FIXED: THE ONE THING THAT IS CERTAIN IS THAT THE OTHER PARTICLE IS ALWAYS SPINNING THE OTHER WAY, EVEN IF THEY ARE LIGHT YEARS APART. IF ONE OF THEM IS FORCED INTO A PARTICULAR STATE, THE OTHER ONE KNOWS ABOUT IT INSTANTLY. BUT ***EINSTEIN*** SAYS THAT'S NOT POSSIBLE – AFTER ALL, NO INFORMATION CAN TRAVEL FASTER THAN LIGHT.

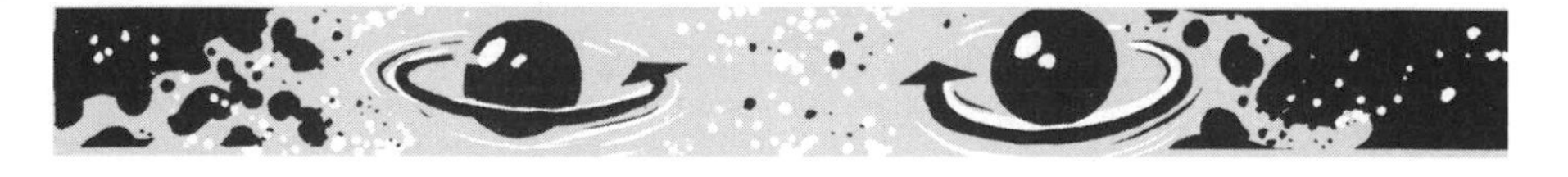

EINSTEIN CAN THINK OF A SOLUTION TO HIS PARADOX: THE PARTICLES HAVE SOME KIND OF INTERNAL MANUAL THAT TELLS THEM WHAT REAL VALUE THEY MUST ADOPT IN ANY GIVEN CIRCUMSTANCE. IT LOOKS AS IF IT'S A MATTER OF CHANCE, BUT NATURE IS STILL NICELY ORGANIZED, DEEP DOWN INSIDE.

IN ***1964***, THE IRISH THEORETICIAN ***JOHN BELL*** THOUGHT UP A TEST FOR EINSTEIN'S PREDICTION. PAIRED PARTICLES ARE EXAMINED USING TWO MEASURING DEVICES THAT THEMSELVES ARE RANDOM IN NATURE. QUANTUM THEORY SAYS THAT THE PARTICLES WILL ONLY TAKE ON NEATLY OPPOSING VALUES IF RANDOM CHANCE MEANS THE MEASURING DEVICES ARE IN THE SAME STATE. ***EINSTEIN***'S IDEA PREDICTS THAT THE OBSERVED VALUES MUST ALWAYS BE OPPOSITE. BUT ***BELL*** FORMULATES A SIMPLE TEST: IF THE RESULT OF THE MEASUREMENTS IS GREATER THAN THREE QUARTERS, WE ARE LIVING IN A QUANTUM WORLD. IF THE RESULT IS SMALLER, ***EINSTEIN*** WAS RIGHT AND QUANTUM THEORY IS SIMPLY NOT GOOD ENOUGH.

EARLY IN THE ***1980***S, THE FRENCHMAN ***ALAIN ASPECT*** PUTS IT TO THE TEST. HE BUILDS AN EXPERIMENTAL SETUP WITH LASERS AND MIRRORS AND HE SEPARATES PAIRS OF PHOTONS. THE TEST SEEMS ***IRREFUTABLE***: EVEN AT A DISTANCE, EACH PARTICLE KNOWS EXACTLY WHAT IS HAPPENING AT RANDOM TO THE OTHER...

THEY ARE LIKE TWO LOVERS WHO CAN TELL INFALLIBLY WHAT THE OTHER IS FEELING, NO MATTER HOW FAR APART THEY MAY BE. ***EINSTEIN***'S 'SPOOKY ACTION AT A DISTANCE' REALLY DOES SEEM TO EXIST.

BUT THERE'S STILL ROOM FOR DOUBT. THE MEASUREMENTS TAKE TOO LONG, THEY'RE TOO COMPLICATED, THE MESH ISN'T FINE ENOUGH - TOO MANY LOOPHOLES IN THE LAWS, PHYSICISTS RECKON. UNTIL THE HOLES ARE PLUGGED, THEY CAN'T BE SURE IF EINSTEIN'S GHOST REALLY EXISTS. SCIENTISTS HAVE NOW BEEN BUSY FOR YEARS TRYING TO INVENT WAYS OF CARRYING OUT THE BELL TEST THAT WILL MAKE IT ABSOLUTELY WATERTIGHT.

ENTER ***RONALD HANSON***, A YOUNG AND AMBITIOUS PHYSICS PROFESSOR IN DELFT, WITH A COUPLE OF YEARS IN ***AMERICA*** ON HIS CV, WHO IS WORKING ON A SUPERCOMPUTER.

AND THERE'S ALSO HIS BRILLIANT YOUNG PROTÉGÉ ***BAS HENSEN***. IN 2013, HE PRODUCES A PROPOSAL FOR FINALLY FINDING A DEFINITIVE ANSWER TO ONE OF THE MOST PRESSING OPEN QUESTIONS IN PHYSICS.

IT ALL STARTS WITH A SKETCH IN ***HANSON***'S ***NOTEBOOK***...

THE SKETCH SHOWS A DIAGRAM THAT MOST PHYSICISTS ARE NOW FAMILIAR WITH: A SETUP WITH A SPINNING PARTICLE AT BOTH ENDS AND A METHOD FOR MEASURING WHICH WAY IT IS SPINNING...

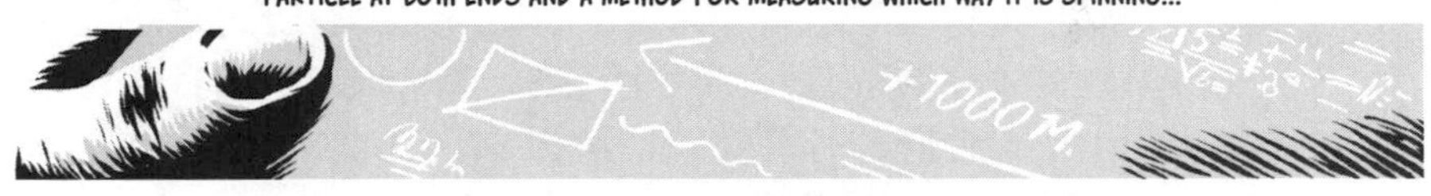

THERE'S ONE THING THAT'S UNUSUAL ABOUT ***HANSON***'S SKETCH, THOUGH: THE ***SCALE***. NORMAL QUANTUM EXPERIMENTS FIT ON A LAB BENCH, BUT THIS ONE HAS TO BE KILOMETERS ACROSS - AS BIG AS THE WHOLE DELFT CAMPUS. THE LARGE DISTANCE IN THE SETUP IS NEEDED SO THAT MEASUREMENTS CAN BE MADE AT BOTH ENDS BEFORE THEY WOULD HAVE TIME TO EXCHANGE ANY SIGNALS.

WE'RE NOW IN THE FALL OF 2014. ***HANSON*** AND ***HENSEN*** TAKE A MAP OF DELFT TECHNICAL UNIVERSITY. USING A COMPASS, THEY DRAW IN A CIRCLE WITH A RADIUS OF ONE KILOMETER AROUND THEIR OWN LAB. WHERE'S A GOOD PLACE TO THE OTHER END OF THEIR ***GIGANTIC*** EXPERIMENT?

THE FIRST CHOICE WAS A HOUSE SCHEDULED FOR DEMOLITION CLOSE TO THE TU. BUT A ***BREAK-IN*** SOON SHOWED HOW VULNERABLE THE LOCATION WAS...

THE BASEMENT OF THE NEARBY DOMED ***NUCLEAR REACTOR***, SAFE BEHIND HIGH FENCING, SEEMED A BETTER OPTION.

TOGETHER WITH TECHNICIANS FROM THE TU, A ROUTE WAS WORKED OUT FOR THE CABLE TUNNELS UNDER THE CAMPUS FOR THE KILOMETERS OF ***FIBER-OPTIC CABLE*** THAT WOULD BE NEEDED FOR THIS EXPERIMENT.

THE ***QUANTUM CONFIGURATIONS*** WERE BUILT IN THE SUMMER OF 2014. THEY ARE PALACES OF LASERS AND MIRRORS ON VIBRATION-FREE TABLES MASSES OF MEASURING EQUIPMENT AND KILOMETERS OF CABLING. IN THE CENTER OF THIS HALL OF MIRRORS, THERE IS A MINISCULE DIAMOND THAT CONTAINS THE CORE OF THIS EXPERIMENT: INSIDE A CAVITY IN THE GLITTERING GEMSTONE, AN ***ELECTRON*** IS SPINNING AROUND ITS AXIS...

BETWEEN CHRISTMAS AND NEW YEAR 2014, FATE STRIKES A BLOW. WHEN THE RESEARCHERS TURN THE SETUP BACK ON AGAIN AFTER THE BREAK, THE PART BY THE REACTOR – KNOWN AS ***BOB*** – REFUSES TO PLAY BALL.

THEY SPEND WEEKS TRYING TO GET THE DIAMOND CHIP WORKING, BUT NOTHING HELPS. JOKINGLY, THEY RECKON THAT ***EINSTEIN***'S GHOST MUST BE SABOTAGING THE TESTS. THE EXPERIMENT IS ONLY ABLE TO START FOR REAL IN MAY, WHEN THEY GET A NEW CHIP.

ON A VISIT TO ***VIENNA, HANSON*** SEES THAT ***ANTON ZEILINGER*** - SEEN BY MANY AS THE OFFICIAL ***'POPE'*** OF ***QUANTUM PHYSICS*** - IS ALSO MAKING PROGRESS WITH HIS OWN BELL TESTS. ***HANSON*** URGES HIS TEAM TO HURRY UP: THE FOREFRONT OF SCIENCE HAS NO PRIZES FOR BEING SECOND.

THEY GET THERE IN MID-JULY. AFTER THREE WEEKS OF MEASUREMENTS, IT IS CLEAR THAT THE QUANTUM PARTICLES RESPOND TO EACH OTHER OVER A DISTANCE OF KILOMETERS. FASTER THAN LIGHT.

THERE IS HARD EVIDENCE: THE GHOST THAT ***EINSTEIN*** FEARED SO MUCH REALLY EXISTS.

BUT NO ONE KNOWS HOW FAR THE COMPETITORS IN VIENNA HAVE GOT. AN ARTICLE MUST BE PUBLISHED! ***BAS HENSEN*** STARTS WRITING FEVERISHLY...

HANSON WANTS THE ARTICLE TO EXCLUDE ALL DOUBT. THE MANUSCRIPT IS TWEAKED AND FINE-TUNED, FINALLY GOING TO NATURE AT THE END OF AUGUST...

DELFT GETS THERE FIRST! ***ANTON ZEILINGER*** SENDS HIS CONGRATULATIONS FROM VIENNA.

MEDIA ANNOUNCE THE BREAKTHROUGH ***WORLDWIDE***.

HENSEN IS ON HONEYMOON IN AUSTRALIA. HE IS DELIGHTED TO SEE REPORTS OF THE ***DUTCH UNIVERSITY*** THAT HAS PROVED, NO MATTER WHAT ***EINSTEIN*** CLAIMED, THAT GOD REALLY DOES PLAY DICE. THE END

LASERS

It was a cold late afternoon in October 2003 when I joined physicist Wim Ubachs at the VU University in Amsterdam. Press releases had informed me that he could make ultraviolet laser pulses in his laboratory that would make the rest of the world sit up and take notice. Ubachs and his team did that by firing intense laser light into a cloud of krypton gas. The intense light rips electrons from the gas atoms, speeds them up on the back of the light waves and then dumps them back into the atoms. And the enormous intensity does strange things with the atoms.

Just like there are high harmonics in a piano's sound if you hit it too hard, krypton atoms can also emit light of a much higher frequency than normal. 'Compressed light' was what the newspaper dubbed this light trickery by the VU researchers, which could prove useful for chip manufacturing by making it possible to etch even smaller structures.

During the tour of the laser lab full of beams of light, mirrors, and lenses that afternoon, I must have accidentally looked into a laser, because two days later something strange started happening to the vision in my left eye. The parallel lines of the Venetian blind were suddenly curved and watching television was an almost surreal experience. A black balloon appeared to be floating continuously in my field of vision.

The next day, an ophthalmologist found a blister on the retina of my left eye, suggesting partial detachment. We would first try to fight it with eye drops to increase the intraocular pressure, but surgery couldn't be ruled out. The problem gradually disappeared. Just as well, because I didn't like the sound of having eye surgery.

That shows the dangerous power of the quantum, as embodied by lasers. Lasers are the prime example of an application of quantum physics that a classical world would never come up with. The basic idea goes back, once again, to Albert Einstein. It was he who, in 1917, thought of what is known as a population inversion, a term for a situation where electrons in an atom are gradually driven up into higher energy levels without having them drop back down immediately. The atom is pumped up with energy until the electrons all drop down at once. The accompanying energy is released as light (or other form of radiation) of precisely one single wavelength or color.

At the time, Einstein made the first quantum statistical calculations of what later became known as stimulated emission. It seemed like he was fooling around and nobody bar a few science fiction authors considered its uses.

That changed around the Second World War, when American physicists began to explore the production of powerful radio waves by stimulated emission. Several labs worked on what became known as masers, in which microwaves caused the inversion. Those closely involved in the research included the Dutchman and later Nobel prize winner Nicholaas Bloembergen, who had fled a devastated Europe after the war. In the decades that followed, stimulated emission was shown to work in the visible spectrum as well. This was initially only in large laboratories with expensive ruby crystals as the source, but they gradually spread to everyday devices such as CD players and printers. An oddity from the quantum world eventually became something ordinary.

7 SPINACH AND MIGRATORY BIRDS

IN WHICH FRAGILE QUANTUM PHENOMENA TURN OUT TO OCCUR NOT ONLY IN ISOLATED LABS, BUT ALSO IN THE NATURAL WORLD, FROM GREEN PLANTS TO GEOMAGNETIC NAVIGATION IN MIGRATORY BIRDS.

It's been on my iPhone for three months now, a video I made of the American quantum physicist Seth Lloyd, who works at MIT in Boston. It was filmed during a guest lecture in Delft. He introduces himself jocularly as the quantum mechanic, the guy who'll fix all your quantum problems. It's one way for a theoretical physicist to maintain his position at a technical university like his. Not that it's all that much of an issue. Physicists and engineers are collaborating worldwide to build the first quantum computers. You need all the help you can get then.

That's true in Delft too, where I interviewed him for the newspaper on a windy autumn day, in the lobby of the hotel where he was staying while he visited his quantum colleagues at the university for a few days. Lloyd turned out to be a small, cheerful man of indeterminate age (actually born in 1960), wearing trainers. He has twinkling eyes, a receding hairline, and a thin ponytail that is sometimes hidden under a leather hat. The following day he gave a lecture that the audience would never forget. There's more than one way to give a lecture about quantum theory.

He had already said in the interview that he would dance at one point during his lecture. As far as he was concerned, it was the best way to illustrate what he wanted to say: how delicate quantum processes are sometimes not ruined by a bustling environment, but instead can be maintained or even promoted. Lloyd had already explained in the interview that quantum phenomena play an essential role in the molecular machinery of life. Without quanta, there would probably be no life at all.

That's an amazing conclusion. As a physicist, you associate a quantum device with one thing above all: vulnerability. There's a very good reason why quantum theory is a theory of atoms and particles, far removed from the daily hustle and bustle. Quantum computers have to be ridiculously well isolated if they're going to work, something that the engineers are nowhere near mastering yet. Quantum cryptography owes its very existence to the fact that quantum states can't be disturbed. Any interference is instantly fatal, because that crucial quantum vagueness unfortunately then collapses.

Life, on the other hand, is all about interactions. It's messy and bustling, warm and motile. A wet soup. Not exactly the environment that quanta are happy in, you'd think.

But it's been made clear to me in the meantime that that's all old hat now. Quantum theory is by no means only about cautious physicists and clever engineers any more. Biologists and chemists have pounced on what now promises to become a fascinating new field, quantum biology.

From a historical point of view, this is a weird reversal. Erwin Schrödinger, physicist and founding father of the wave equations in quantum theory, switched to biology in the 1930s. A field he thought offered more for the future. In 1944, he published a booklet called *What is Life?* that is still considered a classic because he predicted in it that life would have a blueprint that had to involve a code in some kind of 'aperiodic crystal'. Precisely that 'crystal' was discovered and described ten years later, the molecular spiral staircase of DNA. At the time, biology was benefiting significantly from physics. The reverse seems to be taking place now.

I tap my phone screen and show Einstein and Bohr the clip of Seth Lloyd dancing.

The American professor can be seen coming in from the left, with his arms lifted as if he is doing the *sirtaki*, shuffling his feet alternately forward. He is shaking his head frantically. Meanwhile, he tells us that this is the excitation caused by a photon from the sun in the heart of a chloroplast. Then he slows down and stops, as if grinding to a halt in a fixed state after only covering half the route to the end point. He lets his head drop more and more.

He is suddenly given an imaginary push in the side… and there he goes again, shuffling frantically, his arms lifted, until he is back behind his lectern to continue his story about quantum biology. The push in his side, he explains, was the chaotic outside world, which had prevented the system from giving up too soon.

Bohr and Einstein are looking at the screen in disbelief.

The amazing physics will be coming in a moment, but you can already see them thinking that education isn't what it used to be. Too right – more so than they think. Only a few moving images of lectures given by these two men have been preserved, and to be honest, those images give good grounds to fear for the quality of the lecturing at the time. Stiff men with soft voices and thick accents. Hardly interested in their audience. It may have been the camera that made them feel awkward, of course, but it's not exciting at all.

I keep my mouth shut. It would be a pity to create a fuss at the very point that our real quanta will perhaps be at their most tangible. And that's what it was all about, today: showing that, although quantum physics is strange, it is certainly not the exclusive domain of a handful of smart physicists.

Lloyd's story – and quantum biology – started during a working lunch in 2007, when his group discussed a story that had just been published in the *New York Times*. Plants, the paper said, are quantum computers. Much hilarity around the lunch table. If MIT engineers like them found it so extremely difficult to create a functioning quantum computer, how could a plant do it just like that?

The story in the *New York Times* was about experiments by Graham Fleming's group at the University of California, Berkeley. Fundamental research was being carried out there into how chlorophyll works (that's one of the molecules that green plants use to make sugars and oxygen, the process called photosynthesis). Richard Feynman, one of the founding fathers of modern quantum physics, once described it as the process where a photon kicks an oxygen atom out of a carbon dioxide molecule. But it's nowhere near that simple.

For decades, Fleming has been using lasers to study the way that light energy is absorbed and passed inside the chloroplasts of green plants. Understanding the complicated molecular mechanism is just one of the challenges. The biggest question is actually the efficiency of the process. In chlorophyll, light energy is transformed into chemical energy in the form of sugar molecules. There's almost zero energy loss and it's a mystery how this can be possible. Most cellular processes take place with a lot of other extraneous activity too. It works in the end, but it's certainly not efficient. So why is it in chlorophyll?

A small biological detour is required to give us a better understanding of this mystery. Einstein and Bohr are nodding in perfect synchrony. They seem intrigued, as almost all of this is new to them.

But not quite all of it. Around 1900, Einstein himself had thought about random movements at a microscopic scale, in a drop of water, for instance. The molecules in it are moving about at random, continuously. That can even be seen, for example, when a tiny pollen grain is floating in the droplet. You can actually see how this grain, just a fraction of a millimeter across, is pushed and shoved randomly, making it follow an erratic path. This effect had already been observed by the Scottish horticulturist Brown well before that time, although he concluded that the grains clearly seemed to have some form of vitality. But that's not it. The path of the particle that this Brownian movement is subjected to, wrote Einstein in his younger years, is nothing else but the well-known 'random walk', driven by the chaotic movements of water molecules around it. In the random walk problem, also known as the drunkard's walk, the average distance from the starting point depends on the square root of the elapsed time: after four seconds, the drunken man changing direction at random is twice as far away as after one second. After nine seconds, he is three times as far away on average, and so forth. Not really the best way of making progress.

The funny thing is that this effect is also important for photosynthesis. You also first need to understand how chlorophyll absorbs sunlight, so I take a beer mat to sketch that out on the back. The same brand of beer I'd been drinking.

Light is absorbed inside the chloroplasts, which are microscopic organs that float around in the cell plasma of green leaves. Sunlight is white, and plants appear green because they absorb the red part of the spectrum, and reflect the blue and yellow parts. The absorption is what we're interested in.

The chloroplasts are surrounded by a transparent membrane, within which flat complexes of pigment molecules are stacked like miniature molecular checkers. We have now gone beyond the domain where a microscope can see anything, but we need to go further. Here and there, bundles of chlorophyll molecules are popping up like antennas from what are known as thylakoids, the structures that collect light and are the starting point for photosynthesis.

Chlorophyll is an almost flat hydrocarbon molecule with a group of nitrogen atoms on the end, which surround exactly one magnesium atom like a cage. The electrons in the outer shell of the magnesium atom are the photosensitive component of the antenna. They are loosely attached to the rest of the atom in such a way that a well-aimed photon can knock one out of its orbit and into the cage. This has transformed the energy of the photon into the energy involved in separating the negative charge of the electron from the positively charged magnesium ion that remains. Physicists call such an energetic duo an exciton, because it looks a bit like a particle in some respects. In addition to energy, it also has a location. And, even more importantly, it can move.

And that's exactly what is needed. Excitons are not stable; in this case, the positive and negative charge attract each other and will tend to come together. The energy is then transformed into heat and will be lost for further use. In order to prevent that, the exciton must be transported quickly into a molecular complex where the energy of the charge separation is transformed into chemical energy, at the so-called reaction center. The efficiency of photosynthesis depends entirely on the effectiveness of this transport. If this process is messy, the benefits of collecting sunlight will be too low.

In principle, nature is prepared for this. The chloroplasts are packed together so closely that excitons of one chlorophyll molecule can be transferred to one of the surrounding molecules. The real mystery is how they know what the most efficient route is. There are numerous pathways leading to the reaction center, and most of them are so long that the exciton will almost certainly be lost along the way. The random walk that is seen in most chemical processes is utterly inappropriate for this. That can't get excitons to the reaction center.

Einstein is staring at the beer mat. Molecules but no random walk? Remarkable.

At this point we have to return to Graham Fleming, the Californian researcher who suggested in the *New York Times* in 2007 that photosensitive bacteria and green plants are quantum computers. The measurements made by his group at Berkeley showed that the excitons in the chlorophyll-like FMO system of a cyanobacterium really do seem to know the best way from the antenna to the reaction center. Because, said Fleming to the newspaper with a feeling for drama, they try out all possible routes at the same time, and only the quickest one is then taken. It's the traveling salesman problem. Down in the world of molecules and solved with quantum magic that regular quantum engineers can get nowhere near.

Hocus pocus, thinks Seth Lloyd in Boston. But after the cheerful working lunch, the quantum mechanic couldn't get it out of his head, so he started to calculate. He pictured the exciton at the first magnesium atom: a quantum mechanical energy packet that has to go through a maze, more or less like an electron going through a screen with two or more slits. In the famous two-slit experiment we know what happens next. The electron doesn't pass through one of the two slits, but finds its way like a kind of wave through both slits, with interference making it reach some points more easily than others behind the screen. The weirdness of quantum theory starts with the idea of a particle that can interfere with itself.

Lloyd supposed that the same thing was happening in chlorophyll too. The exciton is a wave that tries to find its way to the reaction center via all possible paths, and it materializes there after interference with itself. It doesn't walk about drunkenly. It takes all paths at the same time.

With that revolutionary representation of the situation, Lloyd soon found a formula for the efficiency of chlorophyll, which also predicted how it would depend on the temperature of the system. And when he checked this relationship in the literature, he was perplexed. Chlorophyll does exactly what his calculations showed. Plants really are quantum computers, he concluded, the energy of the collected photons doesn't take a random walk to the reaction center – it's a quantum walk. His earlier assessment of the quantum hocus pocus of Fleming and the newspaper was too quick and easy, he has since acknowledged more than graciously.

When Lloyd later published his results, the responses were not all wholly positive. Fleming's measurements, for instance, were carried out on a chlorophyll complex from a cyanobacterium that had been cooled in liquid nitrogen to almost minus 200 degrees. That made measuring easier, and although it still works, these are clearly not the conditions in which real plants and bacteria normally function. This 'photosynthesis' was probably just the umpteenth quantum effect that could only be observed in isolated labs, and not really important in everyday reality.

That has changed now, I hasten to tell Bohr and Einstein, before they speak. I pick up my list to make sure I won't forget anything. In 2009, a group in Dublin carried out experiments with a photosensitive complex in a bacterium at room temperature, and they observed similar effects. In 2010, the University of Ontario added photosynthesis in algae to the list. Fleming's group had in the meantime found quantum effects in spinach leaves, which have a specific chlorophyll system that occurs in roughly half of all green plants in the world.

The gist of it is that the transfer of the energy package only needs 600 femtoseconds (millionths of a billionth of a second) to reach the center. The idea is that that is so quick that outside interference gets no chance to affect the normal decay of the quantum state. The green quantum plant is just too fast for the disruptive outside world. It's a lesson that quantum engineers like yourselves in particular should never forget, Lloyd tells his audience in the Delft lecture hall. As well as isolation, speed can provide useful protection against decoherence, for instance in the qubits in a quantum computer. Speed could make the difference between quantum computers that need to be in ultra-cooled bunkers and (who knows?) a quantum smartphone in you trouser pocket. Twilight is falling outside.

The story about photosynthesis that Lloyd was depicting with dance for his audience of students and staff that Thursday afternoon is just one of the examples of biological systems in which quantum phenomena play an essential role. The field of quantum biology now extends much further and covers such different items such as the action of enzymes, how migratory birds and tortoises orientate themselves, and there even are indications that our own sense of smell is governed by special quantum effects too.

I can see them looking, Einstein and Bohr. *Die verteufelte Biologie* – that damned biology – has never been able to fascinate them. It's too complicated, and often too *ad hoc* as well. But now that entire organisms seem to use real quanta in their processes, that's certainly going to have to change. Maybe their silly *Herr Kollege* Schrödinger wasn't so completely wrong at all when he switched to biology. And the very suggestion that quantum computer builders could learn something from chlorophyll and spinach, is very exciting, of course.

But that's beside the point. Isn't it maybe time for a snack, after all the words that have been spoken?

For the sake of convenience (as well as being ever so Belgian), all three of us order mussels and chips and a nice glass of beer. And by the time the coffee and chocolates arrive, we're ready for the rest of the story. And even mussels and beer have a pinch of quantum flavor, as will become clear later on.

The story starts off by being all about enzymes, the chemical catalyst compounds that essentially keep the machinery of life running. Without enzymes, biochemistry would hardly exist. We wouldn't be able to breathe or digest our food. Our metabolism would not be able to function without them. In these processes, enzymes are what chemists call catalysts, substances that promote other chemical reactions without being consumed themselves.

Most of the substances that we know are more or less stable. If they weren't stable, we would not be able to see them around us so much; we'd obviously be far more familiar with their reaction products. The stability of a compound can be pictured as the inability of the components to separate from each other. Usually, this is a matter of energy: it's as if the compound has to cross a mountain range to break down into components. In some cases this can for example be achieved by heating the substance. This gives the compound enough energy to get across the high mountains; in fact, the chemical bond will be shaken about so much that the chance of it breaking increases rapidly.

The processes are often more complicated in biology, though. Collagen for instance, a component of connective tissue, is one of the most difficult proteins to break down. Connective tissue gives shape and solidity to the tissues of living creatures; it is literally the elasticity that makes some mussels just a little difficult to chew.

In some fossils of tens of millions years old, it is still

found more or less intact and flexible in rocks, sometimes with the tendons and blood vessels still in it. All other substances are petrified and decayed. But not the connective tissue.

The reason why collagen seems so indestructible is that it consists of long strands of amino acids that are attached to each other with a peptide bond between a carbon atom and a nitrogen atom. The two hold onto each other very tightly because they share electrons and attract the positively charged atomic nuclei on either side towards each other at the same time. It's the electronic superglue of the connecting tissue. To release them, the electrons in question have to be separated. That takes a lot of energy: there really are high mountains between bound and loose amino acids.

These highlands are a sign that the chemical reactions needed for pulling collagen apart are normally very difficult to achieve. For collagen, the requisite reaction takes place via water molecules that join the game temporarily. In principle, that should work. But the intermediate compounds are extremely sensitive to disturbances; they all almost always fall apart again before the reaction has taken place.

This is where a catalyst can help. The chance that the reaction will be completed increases if the intermediate compounds are made more stable. This is, for instance, the reason why some metals function well as catalysts: they provide positively charged ions, which leave negatively charged intermediate products in slightly less of a hurry to get rid of the excess electrons. In terms of the energy landscape, those high mountains suddenly turn out just to be foothills that can be crossed easily after a closer look, thanks to the presence of a metal or other catalyst.

So far, the protein chemistry of collagen (for instance) is more or less classical chemistry, the type that can still be illustrated using ball-and-stick models. It is about atoms and compounds that have to change from one configuration to another. Sometimes they need a little help along the way; a good shake can be enough, or something closer to a headlock may be needed. But where is the quantum component?

It was discovered after the calculations were done. The process where collagen is broken down by a special enzyme called collagenase is understood in some detail. The enzyme contains a zinc atom, for instance, which transfers an electron from one place to another at exactly the right moment in a complicated molecular dance involving the proteins. The process is wonderfully elegant, and all the more impresses if you realize that this solution to the problem of breaking down collagen has been reached through evolution and natural selection. It is of major biological importance. If connecting tissue really was indestructible, growth in animals would be impossible. Or the transformation of tadpoles into frogs. Newts would have tails that could not grow back.

There is only one problem: collagenase does not work efficiently enough if it has to find its way through a confused mass of molecules. Sure, it certainly can speed up the breakdown of collagen a little. But doing so by a factor of billions, which is what it does in practice in living creatures, seems impossible. What collagenase seems to lack in the conventional theory of enzyme processes is effectiveness. Those processes shift electrons and charges about, but it's just all too random.

In exactly the way that energy packets in chlorophyll seem to find their way quantum mechanically through a maze of possibilities, or the way an electron can get through two slits, enzymes like collagenase find a solution too. Such enzymes manipulate electrons in protein molecules. But the ball-and-stick model of reality is then no longer sufficient, and it's time for quantum mechanics. Of itself, that's not a new insight. American researchers discovered back in the 1960s that the electrons stored in respiratory proteins continued to move about, even at temperatures far below zero. The enzymes that handle the electron transport seem not to need heat in order to do their work. In classical terms, that's incomprehensible: if there's any kind of 'highland' blocking the road, the reaction stops.

But not in quantum mechanics. All physics students have at some point tried to get their head around the quantum problem in which a particle comes up against an impassably high barrier. Me too. In classical terms, it's clear what will happen: the particle can't get any further and will always remain on the original side of the wall. But in quantum mechanical terms, a particle is also a probability distribution, a wave. If you do the sums carefully, you'll discover something astonishing: the probability of the particle remaining on the left-hand side of the wall is high, but the chance of it being found behind the wall is not zero. This effect is known as quantum tunneling and it was first described by Friedrich Hund in 1926. And tunneling isn't just a minor idiosyncrasy of quanta. It helps explain the radioactive decay of some atoms. Conversely, the nuclear fusion that allows the sun to radiate light and heat onto the Earth is only possible if hydrogen nuclei are able to get close enough together, against all expectations, for them to be able to fuse and create helium. Some electronic circuits only work because electrons are able to tunnel through a barrier.

That also happens in the case of respiratory proteins. Even at low temperatures, electrons can be shuffled around the place within them because they are regularly able to pop through some energy barriers. For electrons that is perhaps not so utterly miraculous, given that they are pretty much *the* particles that were used when the wave nature of particles was discovered.

More amazing still is the fact that experiments have shown over recent years that entire hydrogen atoms can be shifted from one place to another by tunneling in certain enzyme reactions. Or at least in part. A weakness of the numerous experiments is that they generally show that the quantum effects only start to come into play at very low temperatures. Whether they are even possible at normal biological temperatures has been a topic of heated debate among the experts. Everything in biology is in essence based on atoms, is the usual response, but that doesn't have to mean that all biology is actually atomic physics.

That's often how it goes when discussing quantum effects in biology. The idea that delicate quantum states could play a surprisingly large role in the chaotic, warm confusions of life is so contrary that many people seem primarily to be putting their efforts into debunking the body of evidence that points in that direction. Maybe the clearest example of all is the way in which some birds and marine animals find their way around our planet. Their navigational skills, it is becoming increasingly clear, are an unprecedented piece of practical quantum mechanics.

I give one of the waiters a sly wink. "I reckon," I say to

him, "that we could probably do with a cognac by now. Pick a good one."

Einstein and Bohr agree, wordlessly.

How do migratory birds find their way from the north of Europe to the north of Africa and back again? It's an age-old question, one that has occupied everyone from mystics to sensory physiologists, but for which a genuine answer only seems to have been found in the last decade or so. Quantum mechanics turns out to be involved again.

The bird that has been examined most is one we all know, although I don't think many people realize that it can also be genuinely migratory: the European robin. When winter comes, they leave southwards *en masse* from the far north of Scandinavia, primarily to Southern Europe but sometimes even further, to North Africa. Even when they're making that journey for the first time, they seem to know exactly which way to head.

In the mid-1970s, a German man-and-wife team of researchers discovered that they seem to be able to use the Earth's magnetic field for this in an unusual way. Robins don't seem to have a normal compass like human travelers have been using for centuries. Their ability to orientate themselves depends primarily on the angle that the Earth's magnetic field makes with its surface.

The Earth can be seen as a large bar magnet, with the field lines going in sweeping curves from the North Pole to the South (the same lines that can be made visible for an ordinary magnet by scattering iron filings around it). At every latitude on the planet, those field lines come out of the Earth's surface at a different angle. And these small migratory birds seemed able to sense that in some fashion.

That doesn't sound so strange until we realize just how

weak the Earth's magnetic field actually is – generally somewhere around the 50 microtesla mark. That's enough to make a compass needle rotate, but it is thousands of times more feeble than a common-or-garden fridge magnet. How a signal that weak could play a role in the sensory perception of a living animal remained an enigma until a German called Klaus Schulten discovered at the end of the nineties that a robin also needs light before it is able to detect the magnetic field.

Schulten was the man who, twenty years earlier, had thought up a mechanism by which a living organism could be able to detect an extremely weak magnetic field. To do that, you would need a biomolecule in which radicals could occur, atoms with an unpaired electron in their outer shell. That means that they must have a net spin, a small magnetic quantum moment. On its own, there's no useful way in which that could detect a magnetic field, but there would be if a specific reaction could produce two radicals at once with spins that were quantum mechanically linked. A pair like that, Schulten was able to show, acts like a molecular compass needle. Not in the sense that you could read it, but because it could then make chemical reactions of the molecule in question sensitive to magnetic fields.

It became clear what molecule that was in 1998: cryptochrome, a complex biomolecule that plays a part in the avian eye's sensitivity to light. Schulten dropped his work at the University of Illinois on photosynthesis unceremoniously and reopened the Robin file. The article about the role of quantum entanglement in migratory birds' orientation to magnetic fields appeared in the prestigious journal *Nature* in 2004.

What is striking about it in particular is how natural it

must seem to the bird to be able to perceive a weak magnetic field. They don't 'feel' South – they can see it as a pale patch, because their eyes are just a little bit more sensitive to light in that direction of the Earth's magnetic field than in other direction. Light that hits the cryptochrome molecules in the retina creates the requisite pairs of radicals, which decay more often in the direction of a magnetic field than in other directions and thus increase the sensitivity of cryptochrome to light coming from that angle. The article in *Nature* is seen by many as the first serious start of quantum biology. The esoteric phenomenon of quantum entanglement turned out to give a living organism that we're familiar with in our gardens its superior sense of direction. Even though all the details have not yet been clarified, quantum entanglement exists – as clear as daylight.

Einstein sniffs and shuffles on his chair. "That tale about the robin is intriguing, of course," he says. "But quantum entanglement remains a thoroughly unsatisfying phenomenon, because of that spooky link between two separated objects." He wonders out loud that it may work as a solution for the puzzle with the robins, but deep down, physics does not actually know what is going on. Bohr puts a hand on his arm. "Why does it matter?" he asks calmly. "As long as it works."

That is the pragmatic attitude that got him so far in life. But he realizes better than Einstein that what works one minute can be superseded the next. In quantum biology, that seems to apply in particular to our sense of smell. It was thought for a long time that the olfactory organ, high up in our noses and not so coincidentally close to the brain, had a series of sensors that function roughly like a series of padlocks, into which molecules of various volatile compounds can fit like keys.

There's one problem with that idea though, despite the fact that there were even Nobel Prizes awarded for it in the past: molecules with totally different structures seem capable of producing similar smell sensations, and conversely there are structures that are at first sight identical but yet smell very different. It is a nightmare for any analytically-minded parfumier, making the creation of scents more of an art than a science.

However, there has been an alternative theory about the perception of smells since the 1920s, the time when physicists such as Pauli, Schrödinger, Heisenberg and others were coining the new currency of quantum mechanics. That theory was thought up by the British chemist Malcolm Dyson, who strikingly enough survived German attacks with mustard gas during the First World War. That's one smell you could never forget. But it was less clear why the alternative fitted. According to Dyson, our sense of smell does not respond to the shape of volatile molecules so much as to the way in which they vibrate. It's not something anyone thought about much during his time, but the atoms in molecules do not really have fixed positions. It's more as if they are connected together with tiny springs, not the sticks we're used to thinking of. The possible modes of vibration in the system, physicists later discovered, even constitute a kind of unique fingerprint for the molecule in question. That effect is still happily utilized today in what is known as Raman spectroscopy.

But certainly in Dyson's time, it seemed a bit far-fetched to think that our noses were capable of picking up the vibrations in molecules. On top of that, the theory had its problems too. There are some molecules that exist in two versions that are mirror images of one another. Their vibrations are identical. But they can smell completely different. A well-known example is limonene, which gives oranges and lemons their characteristic citrus smell. The mirror image is called dipentene, and it smells of turpentine.

It seems as if the true answer to the puzzle of smell must lie in a combination of molecular vibrations and sensitivity to shape. Shape seems primarily to determine which receptors in the nose a certain molecule can act on. But what the molecules then do there seems more to resemble pure quantum mechanics. Once it is bonded onto a receptor, a scent molecule seems to create some kind of quantum short circuit between parts of the receptor that would otherwise not affect one another. What seems to be happening, theoreticians have come to think over recent years, is that electrons are tunneling from one place to another in the receptor, in the same way as sound can travel through walls. The proximity of the vibrating odor molecules lowers the energy barrier that would normally forbid this. And the right vibration frequency is an essential aspect. But as soon as an electron in the receptor has made the jump, a biochemical mechanism rolls into motion that ultimately triggers nerves and brain cells and gives us the sensation of an aroma.

Experiments with mice have shown that at least part of this scenario must be right, at any rate. Fruit flies can follow certain scent trails with ease in a maze, as long as they don't come across a compound in which the molecule has a characteristic vibration at around 66 terahertz. At that moment, they turn around in disgust and pick a different route. So far, quantum tunneling is the only plausible explanation for this.

The cognac that I just ordered has arrived. At the table, Einstein and Bohr are now looking pretty tired. The days when they were able to spend all night squabbling and bickering in the name of science are clearly already nearly a century ago.

8 THE EVERYDAY RIDDLE

HOW QUANTUM PHENOMENA ALSO EXIST IN THE VISIBLE WORLD, ALTHOUGH THEY ARE RARELY PROMINENT, AND WHY THAT MEANS QUANTUM IS NORMAL AND THE CLASSICAL EVERYDAY WORLD IS THE WEIRD ONE.

It's nearly midnight. I tell Bohr and Einstein about the time in Leiden that I went round to see Dirk Bouwmeester, an ever-cautious and soft-spoken quantum physicist who also works in Santa Barbara, California. It was the day after he had accepted a prestigious Dutch academic award, the Spinoza Prize, from the hands of King Willem-Alexander, and he was clearly still a bit bleary from all the festivities. He spends a couple of months every year in Leiden, the city where he once studied physics and where his fascination with the quantum also began.

Bouwmeester, who looks younger than he is, is now one of the world's leaders in the field. A quantum grandmaster, who as a postdoc with Anton Zeilinger in Vienna was once the first person to teleport an excited atom from one corner of the lab to the other. *Beam me up, Scotty!*

Who is Scotty, I see the two of them thinking. I let it pass. The next bit will be of more interest to Einstein in particular.

For his intermezzi in Leiden, Bouwmeester has his own *pied à terre* in Witte Rozenstraat in the center. The address is no coincidence, of course. The house diagonally opposite is the former villa of the theoretical physicist Paul Ehrenfest, the tragic hero of Leiden physics, the successor to Lorentz, a friend of Einstein and Bohr and numerous others, as well as the legendary tutor for younger heroes such as Casimir, Goudsmit and Uhlenbeck. But this is also the same Paul Ehrenfest who took his own life in 1933 in Vondelpark in Amsterdam with a pistol. Shortly before, he had shot dead his disabled son Wassik.

"*Dieser armer Kerl*," sighs Einstein. "Poor man." He had often stayed with Ehrenfest in the house in Leiden. He had a violin there, for impromptu moments. He had given a speech at Ehrenfest's funeral.

We sit there in silence for a moment, Bohr, Einstein, and me.

Bouwmeester's office nowadays is in the Huygens Laboratory on the northern periphery of the city of Leiden. Well, laboratory is a grand word for it. The Huygens is basically a flat with a few rooms for working in. Quite whether experiments are being done, or where, is pretty unclear. It could be behind any door on the corridor.

Bouwmeester and his team do their experiments in the low-rise block next to building, where vibration-free floors cut influences from outside down as far as possible. But in fact he's waiting for a new lab that is being built a little further along on the university premises. As irony would have it, the pile-driving for that new building banished his experiments to the nocturnal hours. This new lab is where it is all going to happen.

This is where he wants to conduct a fiendishly awkward experiment to show definitively that there is no boundary in principle between the quantum world and the large-scale world. The idea for it came from the venerable British physicist Roger Penrose, with whom he worked for a number of years in Cambridge. I also spoke to Penrose once myself, at a lecture in Groningen when he had just written a book about thinking. According to him, that takes place through molecular threads in the brain that are able to harvest quantum effects. He's not the man for the minor insights, you could safely say.

There's nothing minor either about the plan that Bouwmeester is currently working on, which is distinctly in the Penrose mold. His question is whether the laws of quantum theory also apply to objects that are much bigger than an atom.

His experiment is perfectly feasible on paper. Bouwmeester wants to show that not only particles but also macroscopic objects can exist in quantum superpositions. It is obvious enough for an electron, for instance. The particle has a spin of a half, that's definite. But whether the spin is up or down is not. When the measurement is made, one of the two possibilities becomes reality; which one is purely random, according to prevailing quantum opinion. But what about bigger objects? Everyday reality does not give you the impression that objects regularly exist in multiple states. But the transition to the domain where that is the case must occur somewhere. Does it happen with two particles? With a thousand particles? With molecules? Tennis balls? Entire planets?

This is actually the question that Erwin Schrödinger, the man of the quantum wave equations, was focusing on with his famous story about the cat. There are numerous versions of the tale, but they all come down to a box containing a cat and a poison gas vial that is broken open if a radioactive atom decays. The decay of atoms is a quantum process that is controlled purely by chance. However, inside the closed box, that random event is a matter of life and death for the cat. Schrödinger was particularly interested in the question of what we can say about the contents of the box if we don't look inside it. In principle, he said, it's not just that the atom exists as a superposition of decayed and unchanged states. As a result, the cat too has to be in a superposition of alive and dead states. No matter how large and solid that cat may be, it is essentially also a quantum system, of which one of the possible states becomes reality at the moment we look inside the box.

Schrödinger's parable of the quantum cat seems to suggest that macroscopic objects, or even living creatures in this case, can exhibit quantum behavior. He didn't really mean it literally, and Schrödinger's cat is more about drawing attention to another thorny problem in quantum physics, the measurement problem.

The measurement problem goes to the very heart of what quantum mechanics is actually trying to tell us. What happens when the spin of an electron is measured? What makes the undefined superposition of spin-up and spin-down crystallize into one of the two when a measurement is done? Has the measurement inevitably – bluntly but clearly inevitably – disturbed the quantum's blissful ignorance? Or is it more like some kind of fundamental lottery, a lucky dip of possibilities with the measurement being what you pick at random with your eyes closed?

It doesn't bother most quantum physicists too much, but they do have a smooth answer to the measurement problem if they need one. The measurement on a quantum system, they then say, associates that system with the measuring device, which is essentially also a quantum system. The rules of quantum mechanics then let you calculate very precisely that there are specific probabilities for the entire thing taking on various states. In jargon terms that's called the collapse of the wave function, an image that does indeed have something going for it: the quantum system was at first like an ocean full of waves, with the measurement acting like a harbor mouth that only allows a single one through easily.

But outsiders like myself get a rather uncomfortable feeling about the collapsing wave function, because it's really a question of semantics – replacing the word 'measure' with the word 'associate' still doesn't solve the puzzle, of course.

One person who did try to do that, I remember because I talked to him at length for the newspaper once, was the mathematical physicist Klaas Landsman from Nijmegen. Together with a student, he made a new calculation a couple of years ago of the measurement problem and noted that there is a thermodynamic reason why a quantum system that is disturbed shifts into a single measurable state. It's a bit like having an egg box with marbles on those two points in the middle. If the box gets a knock, the odds of the marbles falling into one of the dips are vastly greater than the odds of them staying up on the peaks. Landsman was photographed for the occasion sitting on the sofa at home with his two cats. But if we're being honest, we never heard a great deal more about his two substantial articles that claimed to be the solution to the measurement problem.

Back to Leiden. In Bouwmeester's experimental setup, the behavior of a tiny mirror is key. In human terms it is a very small object, weighing just a couple of micrograms, on a feeble nano-scale spring. But it's vastly bigger than anything we expect quantum behavior from. The question is whether this mirror can both vibrate and not vibrate at the same time.

The tiny mirrors are made in a nanotech lab in California; Bouwmeester brings them back himself every now and again in his briefcase. In Leiden, they're working on the measurements. The idea is that a single photon in the measurement setup first goes through a half-silvered mirror. If the photon passes through the mirror, it goes into a tiny hollow where it gives the tiny vibrating mirror a nudge and is then reflected into a detector. However, if the photon conversely does not pass through the first half-silvered mirror, it goes directly to the detector.

The quantum description of the situation is simple: the photon takes both paths and produces an interference pattern with itself at the detector. The experiment is attempting to see whether the vibrating mirror then acts classically, or as a quantum system that is neither only vibrating nor only stationary, but both at the same time.

Bouwmeester's experiment is a bit of a gamble, because he genuinely doesn't know what the result will be. If a tiny mirror weighing a couple of micrograms also turns out to be a quantum system, that obviously means that quantum laws are not so far from reality after all. The question is whether there is even any fundamental difference between classical reality and quantum reality.

That sounds nuts, and it is. How can the indeterminacy of quantum laws fit in with the predictability of everyday reality? The simplest reasoning sounds like pure science fiction... with the emphasis on the 'fiction': every quantum decision sees reality split into different paths in which every time one of the options is taken and the other is not. It's as if we are driving through town and both turning right and going straight on at every junction. Two versions of our lives are literally created: a story in which we went straight on, and a different tale where we chose to go right. And in each version, we don't have the faintest idea of the existence of the other variant.

Bouwmeester has an Israeli colleague, a physicist by the name of Lev Vaidman, who is absolutely convinced about this constant splitting of reality. As a joke, he developed a kind of wristwatch that shows ones and zeroes, determined by spin measurements on electrons. For the important decisions in his life, Vaidman systematically consults this watch to see if he should do things or not. It is a modest attempt to capture the hectic nature of the quantum occasionally, which in reality determines our lives at every single moment, whether we notice it or not.

This many-worlds interpretation, proposed by the American John Wheeler and others, is controversial. Many in the world of physics think it is a cheap way out, believing that the proponents are giving up too easily. Physics can't be a system in which anything goes; reality cannot be a cascade of random events. That's not what we're used to in physics.

That idea always makes me think of the theoretical physicist Max Tegmark, of Swedish origins but now an American, who thought up the ultimate quantum experiment a couple of years back to test whether there is one single reality or many. Take a setup with a half-silvered mirror, he said, and fire a photon at it. Behind the mirror is a detector that is connected up to a machine gun. The gun is pointed at Max Tegmark. The first photon hits the mirror, but there's no bang. Tegmark observes that he is still alive, so the photon was evidently reflected. The same thing happens with the second photon. And the third.

Has Tegmark been lucky? It's possible. In theory he could keep escaping death forever, but the probability is minimal. Sooner or later, a photon has to get through the mirror, hitting the detector and killing the person who thought up the experiment. If that does not happen for a long time, there is only one serious conclusion: for each photon, there is a Tegmark who was shot dead and a Tegmark who survived. The same thing happens with the next photon. If Tegmark remains alive, then he knows that he is in the storyline of all Tegmarks who keep not getting killed. And that storyline is the story of the many-worlds interpretation.

And according to Max Tegmark himself, there is just one reason why he has never carried out the experiment with the quantum machine gun: he has a small daughter and he wouldn't want her to have to cope with having her father commit suicide. Even if it was crucial for the interests of science.

Tegmark is incidentally by no means the only person interested in experimental evidence for the many-worlds interpretation. Previously, for example, John Wheeler and even the mildly wacky theologian and physicist Frank Tipler had made plans for tests. And Lev Vaidman in Tel Aviv has an app on his website that gives a random response to a question after a quantum measurement in his lab. If you type in 'stay' and 'go' and click a dice, you immediately get to see which version of reality you are in. Should I stay or should I go? It says 'stay'.

The less potentially fatal experiments with mirrors in Leiden haven't really been progressing for the past few years, by the way. Bouwmeester is looking for an effect that is so close to the limits of what is observable that any disturbance from outside is immediately catastrophic. Over recent years, the disturbance came from the incessant building work in the vicinity of the Huygens Laboratory. But even now that this has largely come to an end, the torture continues. The micrometer-scale mirrors have now been made, in the lab in Santa Barbara. However, the ultra-cool measurement apparatus can only do its work in an ultra-high vacuum, whereas all the mechanical pumps known on the market turn out to cause far too many vibrations. There are alternatives, but the proof of the pudding is in the eating. It's not quantum theory, but it's certainly a strange field: high vacuum technology.

That's how it always goes in experimental physics, explained Bouwmeester at our most recent meeting in Leiden, the day after the royal handshake. We were sitting in an empty office nine floors up, two guys sitting on chairs and nothing that could disturb them apart from that. Now that he had a Spinoza Prize, this was to be his new working area, but the furnishings would take a while to arrive yet.

Actually, Bouwmeester said thoughtfully, he was getting more and more convinced that we really do live in a quantum world and that the classical, predictable nature of everyday life is in fact an illusion. That we don't experience it that way is just a question of perspective: our own stories are always consistent, but we don't realize that there are an infinity of other possibilities. Ones in which we didn't turn right.

STAINED GLASS

It's quite hard to find them. St. Michael's Cathedral in Brussels is supposed to have a number of hotly disputed stained glass windows called the Sacrament of Miracle. Ten windows in the chapel to the left of the choir that depict the story of the theft of the host from the church on Molenberg on Good Friday, 1370. Six Jews from Brussels stole the wafers and pierced them through, according to legend, in order to make Jesus suffer more. This desecration worked even better than they'd hoped. Much to their surprise, the pierced wafers of the host started bleeding. The bewildered perpetrators were arrested and executed and their families murdered. In 1402 the bleeding 'body of Christ' was officially declared a miracle and a chapel was dedicated to it. Nobody wonders these days whether the story about the Jewish thieves might be true but it was only called into doubt three centuries later.

It's deathly quiet there this Friday, one kilometer from the Hotel Métropole. Sunlight streams in through the colorful windows and that is exactly what I came for, the stained glass. Not because of the undoubtedly concocted story of the Jewish host-stealers, but because of the real quantum story of stained glass, which seems to date from 1861.

That story is in the colors of stained glass windows, which are not painted on but made by mixing metal compounds into the liquid glass. This technique dates back to before the Middle Ages and was no doubt discovered by trial and error. Nowadays we know what's involved in the crimson or cobalt blue glass of, for example, this Sacrament Chapel in Brussels.

For the red, a tiny amount of gold is mixed into the naturally transparent glass, dispersed into microscopic globules. Those tiny beads are so small that electrons on the surface can only move in coherent waves, like overlapping ripples in a bathtub. These plasmons, as they are known, are quantum phenomena that can only occur in certain states, which are determined by the metallic element and the size of the beads.

This produces color effects because light is an electromagnetic wave that influences how the electrical charges slosh around the gold globule. Some colors of the white sunlight that hits it are absorbed perfectly; the ones that are not are transmitted and eventually reach our eyes. White light minus red becomes blue, white light without blue becomes red. We then see blue. Or red.

The color of the glass is what's left of sunlight dancing around tiny gold spheres that are just millionths of a millimeter across. That is the true miracle of this chapel of the Sacrament of Miracle in Brussels.

Incidentally, the very same quantum theory that explains stained glass also explains why metals like gold are so shiny. What is known as the plasma frequency can be calculated for the color globules in the glass, depending on the size and material. The system can absorb light of frequencies lower than this typical frequency, but not higher.

The plasma frequency of any piece of metal more than a few nanometers across is somewhere in the ultraviolet, at much higher frequencies than visible light. In other words, all visible light that hits it is immediately reflected. The electrons on the surface work together to make this happen. So even the glitter of a golden ring is a quantum phenomenon.

Ones in which we didn't coincidentally hold that particular door open for the woman who we would later marry. In which we are not sitting on a chair in an empty room and talking to a journalist about quantum reality.

The real riddle, according to him, is why in the world in each of these quantum narratives can turn out to be so utterly classical. Planets go around the sun, apples fall, some times it rains, an ambulance rushes by with the Doppler effect first raising and then lowering the pitch of its siren. All very classical, so much so that it has already taken a century for us to understand anything about the deeper quantum reality.

Back to Brussels. In the Métropole brasserie, the waiter spontaneously brings us the bill for what we've eaten and drunk during the afternoon and evening, looking pointedly at his watch. "Gentlemen?" What he means is that it's time to go.

The handwritten bill is surprisingly modest for three people over the course of about ten hours. Some coffee, two cups of tea, one Brussels waffle without whipped cream, one *moules frites salade* with a Belgian beer, some more coffee, and a cognac. Where's all the rest?

When I look up, all I see is the waiter in a virtually deserted brasserie. The bar is being cleared up. I pay, take my coat, walked through the hall of the listed building, through the rotating door and out on down the steps. It is misty in Brussels.

A little later in the tram, I check the recordings on my iPhone. I can hear myself talking at length out loud about quanta of light, electrons, atoms, states and matrices, with the sounds of a cafe in the background. About the double slit experiment, about spin, Fermi-Dirac statistics, about entanglement, and spooky action at a vast distance. I go on about qubits in quantum computers, about spintronics, about the quantum compasses in robins, the olfactory cells in our noses, Schrödinger's cat, the Majorana particle in Delft, and Einstein's dicing deity. And about what a big mistake I think it is to maintain that quantum phenomena only occur in remote laboratories and are not important for everyday life. Physicists, I hear myself explain a number of times, really ought to stop perpetuating the idea that quantum theory is too weird to understand. Quanta are strange, to be sure, but it isn't black magic. I'd prefer to say it's the most beautiful thing physics has ever discovered.

If there's one thing that *is* strange, the quantum grandmaster Dirk Bouwmeester taught me in Leiden, it's the everyday classical world. The real mystery is why it is so predictable and unambiguous.

Einstein and Bohr seem not to have said a single word all that time. As if, I realize for the first time, they hadn't really been there at all.

EPILOG

THE DEATH OF SCHRÖDINGER'S CAT

A COUPLE OF WEEKS AFTER THE FIRST EDITION OF 'REAL QUANTA' HIT THE BOOKSHOPS, I FOUND MYSELF WRITING A LETTER TO ALBERT EINSTEIN. IT WAS ALREADY NEEDED: THERE WERE NEW SCIENTIFIC DEVELOPMENTS THAT I WANTED TO UPDATE HIM ON, DEVELOPMENTS THAT IN FACT MEAN HE DESERVED EXTRA CONGRATULATIONS. THAT LETTER GOES AS FOLLOWS.

Amsterdam, July 2015

Dear Professor Einstein, lieber Albert,

This year, 2015, sees the centenary of your general theory of relativity and I cannot congratulate you enough on that more than merely superlative achievement.

Not only because your ideas are still well and truly alive and kicking a hundred years later – how space and time are interrelated and that curvature that we used so unsuspectingly to call 'gravity'. It is crucial, in fact, for anyone who wants to understand how the universe works and how its lifespan has progressed through those billions of years. Getting to know your theory is still a breathtaking experience. How dizzying it must have been to actually think it up for the first time!

Nevertheless, the centenary celebrations for general relativity are not the reason why I am writing to you. I have recently published my book Real Quanta, partly as a result of our meeting – as pleasurable as it was imaginary – in Brussels at the Hotel Métropole that you know so well, in the company of Niels Bohr as well.

My book, about the idea that quantum phenomena are more commonplace than we tend to think, had only just appeared when I read an article in Nature – one of the leading scientific journals in the world nowadays – that will undoubtedly interest you. Even if it was only because it finds a very unusual and surprising way of bringing together two topics that are particularly close to your heart: gravity and quantum reality.

You no doubt remember our discussions in which I explained how much contemporary science is still struggling with the question of what quantum physics is actually telling us. As a toolkit, this branch of physics has proved to be fantastic. We can use it to understand matter at the very deepest level. What is more, a whole world of applications has been created. In chemistry, in technology and even in biology, our ability to describe and predict quantum behavior is playing a massive role.

At the same time, there is still uncertainty about the question of whether our theories are also telling us exactly what is really going on in the world of the very smallest particles. Is everything really so vague that level? Or are we unable to get a real grasp of things because our tools are too blunt?

The last issue, as I told you earlier, is what is known in quantum *physics as the measurement problem: the fact that a particle can only be in multiple quantum states at the same time if it is totally isolated, and that the measurement breaks that isolation. The question here is not even about whether a physical disturbance occurs; the disruption is a more deep-rooted principle, boiling down to the fact that the measurement makes the quantum states of the particle and the entire measurement apparatus merge to create a new quantum state. There is no way of making a measurement unnoticed.*

I was in Zürich a little while back, and I rang the door of number 30 on Hüttenstrasse. *That's right, the old address of your late friend and colleague* Erwin Schrödinger *who was a professor at the university in the 1920s, your university. I had read that there was a cast iron silhouette profile of* Schrödinger's *cat in the garden, an image that could look alive or dead, depending on the angle of the light. Unfortunately the garden of that yellow, listed* fin-de-siècle *building was not open to the public, and the residents had also clearly had enough of my kind of nerdy curiosity. But it was naturally no coincidence that the tale of* Schrödinger's *cat also played a major role in my book about quantum reality.*

You are familiar with the thought experiment of course, although I do not know if it was seen as the archetype of a quantum puzzle in your time in the way that it is today. But I will summarize it now just in case.

We take a closed box containing a radioactive atom and a Geiger counter. When the atom decays, the counter is triggered and a small hammer will break open a vial of cyanide. The box also contains a cat. It was alive when we put it in there, but once the box has been closed we are left with an interesting uncertainty. If we do not know whether the atom has decayed or not, it is in quantum mechanical terms in two states at once, both decayed and unchanged. But in that case, the same applies to the cat. It is therefore both alive and dead at the same time. That is striking, because a cat consists of countless tril-

lions of atoms and you then have to question whether that can still be called a quantum system. Yes, each atom is a quantum object, but their interplay to create a cat is at most a confusing mess you can't make head or tail of (sic). Nevertheless, if we want to believe quantum theory, the quantum cat is a superposition of two states: alive and dead. We observe one of those two states as soon as we lift the lid of the box and look at the cat. Either it is alive or it isn't.

In my book, I described how there have already been countless attempts to understand to what point quantum vagueness can still exist, and where the confusion takes over. The answer is unclear, even yet, even though experiments with large molecules of a couple of hundred atoms have shown that they still exhibit quantum behavior.

And now along comes Igor Pikovski, a young postdoc researcher at the universities of Vienna and Harvard who has added a radical new twist to the tale (not the tail) of the quantum cat. He took the first steps in his thesis last year, although nobody really noticed. But a small storm of scientific excitement broke loose in June, when he published a splendid article in Nature Physics about the question of whether Schrödinger's cat is really a quantum object, or only a question of a closed box and the observer not knowing. Nature put a very apposite header above a comment: "Gravity killed the Quantum Cat."

So gravity has spelled the end for the quantum cat, but what is actually happening? According to Pikovski, there is something that physics has overlooked so far when analyzing larger quantum systems: the time factor. If all the parts of the system are to exist as a cohesive whole, their clocks have to be synchronized. Physicists assume that blindly and don't even include time in their calculations. Pikovski showed however that this is incorrect if the components of a system are widely separated.

According to your very own theory of relativity, not only do moving clocks no longer stay synchronized by definition, but gravity also distorts time. Not by much, but this effect has now been clearly dem-

onstrated. Nowadays there are atomic clocks that drift measurably apart as soon as one clock is about five centimeters higher up in the Earth's gravitational field than the other. And for quantum cohesion to be disrupted, Pikovski showed that a difference in height of maybe a micrometer could be enough.

In other words, the top of a grain of sand on a beach on Earth is already too far away from the bottom of the grain for it to exhibit serious quantum behavior. Let alone a whole cat. Most of the objects around us are simply too big to maintain quantum cohesion, however much they do their best to retain that unity. And that is not a question of disruptive influences, but of the curvature of space that makes clocks drift apart.

The first physicists who I spoke to about this for a story in the paper were gobsmacked. Gravity is quite literally a force that is all about large objects, from grains of sand to planets, solar systems, and even the universe as a whole perhaps. How can that keep the quantum world at bay, if particles and all the other small things in the universe do not even feel gravity?

The answer is the curvature of space that you were the first to calculate, a century ago, after ten years wrestling with the requisite math and the old ideas about velocities, accelerations, and forces. I read in the history books how you, emaciated and dazed from the efforts to calculate the orbit of Mercury, wrote down your all-encompassing formula on the blackboard in the Kaiser Wilhelm Institute in Berlin. From that point on, space and time were a single whole, in which the stars and planets make dents like a bowling ball on a springy mattress so that light does not travel in straight lines near them but is deviated slightly.

I can hear you thinking, "Could it be true? Has the young physicist Pikovski from Vienna got his sums right? And is the predicted effect big enough to explain the absence of macroscopic quantum states?"

I interviewed the quantum thinker and physicist Dennis Dieks from Utrecht, who is also an emeritus these days, about his opinions. And what he had to say is perhaps useful to state here: the influence of time dilation due to gravity over distances greater than a micrometer seems to be greater than you would have expected, but the question is whether it is all that relevant. The fact remains that an object like a cat or a ping-pong ball, or even a grain of sand, will never be isolated in practice in the way that a quantum particle is. There is an overwhelming tsunami of disruptive factors, ranging from heat to inquisitive physicists, that will always find a way of disrupting any superpositions. Perhaps, said Dieks, that is actually more important than the time difference between the quantum cat's tail and whiskers.

On top of that, Dieks said, there is also the question of what Schrödinger's tale from 1935 about the quantum cat teaches us anyway.

It's more about the principle than about the real reason why we never see superpositions of dead and living cats in our lives. But for physics itself, it certainly can have interesting consequences. I also spoke to the quantum physicist Dirk Bouwmeester, who works in Leiden and California, about the work of Pikovski – who turned out to be one of his former students and who even worked on the experiments Bouwmeester is carrying out in Leiden. I told you about those experiments earlier. That is the story of those nano-scale mirrors that Bouwmeester wants to demonstrate can exist in multiple quantum states at the same time. To us, those mirrors are absolutely miniscule, but from the quantum perspective they are gigantic and heavy.

The tiny mirrors that the researchers in Leiden are looking at are reckoned to be small enough not to be affected by the relativistic time differences caused by gravity on Earth. Nevertheless, Bouwmeester admits, it may well be advisable to carry out the experiments in a single horizontal plane as far as possible, so the gravity is at any rate the same for all the components and states.

At the same time, it was clear that the heart of Dirk Bouwmeester, an experimenter pur sang and a justified winner of the Spinoza Prize in the Netherlands, was starting to pump faster. If relativity could indeed have a serious effect on the quantum cohesion of the system, that should also naturally be measurable. Bouwmeester did not say it in so many words, but I reckon he would not have any objection to being the first person to test Pikovski's calculations experimentally. Who knows? Maybe the triumph of gravity over quantum vagueness could start in Leiden, a place you yourself were familiar with in the time of Lorentz and Ehrenfest. In fact, Bouwmeester has a house diagonally opposite Ehrenfest's, in Witte Rozenstraat. Apart from all the cars in the street, you would probably not even think it had changed all that much, were you still alive and prepared to visit Leiden once again.

All in all, I think it's a nice development, that your general theory of relativity now seems to explain why we see so little lower-level quantum vagueness in the everyday world around us. The world we know is simply too big for that, literally. That is not only a neat and really quite surprising insight into physics, but I would say that it in some strange way also justifies your own stubbornness in the entire quantum discussion, for instance with Niels Bohr.

You stuck rigidly to your guns in that debate, maintaining that God does not play dice, that order and causality simply must predominate at some deeper level, that quantum theory was therefore not complete but just the first stumbling steps towards something better and more universal.

I think that physics is in the meantime entitled to say that this opinion is simply incorrect; at the very smallest levels, the world really is made up of vagueness and possibilities. A world of ghosts, whether we like it or not. It can be put more strongly than that: even in everyday reality, some of those shadows can be glimpsed, as I have already told you.

So, no matter how difficult I find it to keep telling you that, you were wrong to say that quantum vagueness is an illusion. But I have understood much better since Pikovski's article in Nature why you *were always so attached to cause and effect and unambiguity, which seem to be the glue holding together most of what happens around us that we can see. Your physical intuition had a classical flavor to it, formed by a world of ricocheting balls, circling planets, ticking clocks, and light signals. You thought about it all more carefully and above all more strictly than all your predecessors put together, and discovered relativity.*

You were of course unable to know at the time that the ultimate consequence of that very same relativity is that quantum vagueness cannot exist at larger scales. More than that, almost nobody had realized that until now. But it does make your certainty that the vagueness could at most be an illusion entirely understandable. In the world that you knew and understood, that kind of elusiveness is genuinely impossible. As a matter of principle, indeed. And you had no way of knowing that vagueness really is the order of the day at the smaller levels, somewhere down beyond the micrometers.

I hope very much to be able to hear your own thoughts about this topic one day.

Your humble and admiring servant,
as always,

Martijn van Calmthout

GLOSSARY

Albert Einstein – German physicist (1879-1955) and Nobel Prize winner, founder of relativity theory and numerous other theories in physics. One of the founding fathers of quantum theory, albeit always critical of quantum phenomena that deviated from the everyday. God does not play dice, he concluded.

Atom – Building block of all tangible matter. Consists of a positively charged atomic nucleus and a number of negatively charged electrons moving around it.

Band gap – The energy jump that electrons have to make in a metal or semiconductor before they are able to move through the material.

Bell's theorem – Mathematical proof that variables in a quantum theory are able to keep sensing each other at a distance. Named after John Bell. Proved experimentally in the 1980s by Alain Aspect.

Binary – The numeric system in which numbers are expressed using powers of two, instead of in powers of ten as in the decimal system. It is used in computer memories.

Bit – A memory element in a computer that can take the values one or zero. It is generally a magnetic element that can be on or off.

Classical – The term used for theories of physics that date back before quantum theory and that generally fit nicely with everyday experience.

Closed shell – An energy level in an atom in which all the possible states of the electrons are filled.

Collapse of the wave function – The effect by which a measurement on a quantum mechanical system only yields one single state from the range of possible states that had existed until then.

Conduction band – Energy state in a metal or superconductor in which electrons are free to move and thereby produce a current.

Copenhagen interpretation – The standard interpretation of quantum theory, which assumes that only measurable variables have any meaning. Its founding father was Niels Bohr.

Cryptography – Secret communication using keys that allow a

message to be hashed and then decrypted again.

De Broglie theory – The idea that a particle is a wave, with a wavelength that depends on its speed.

Double slit – Experiment with a screen that has two slits in it that can be used to show that quantum particles are only located at a single point when they are measured, being spread through space before that.

Einstein-Podolski-Rosen paradox – A thought experiment in which a pair of entangled particles are used for communication over large distances without any time lag. According to Einstein, this was the reason why quantum theory could not be physically complete.

Electron – A negatively charged fundamental particle. It is a component of atoms, and its freedom to move inside materials determines their electrical and optical properties.

Entanglement – A relationship between the values of a quantum property of a pair of particles that continues to exist even after they are separated.

Enzymes – Biological molecules that assist a chemical reaction without actually getting used up. Quantum phenomena such as tunneling play a role in this.

Fermi level – The highest energy level in a metal where electrons are free to move.

Geomagnetic navigation – The ability of some migratory birds to use the Earth's magnetic field as they fly to orient themselves. Quantum effects give it the necessary sensitivity.

Hidden variables – Quantum entities that cannot be measured, but which yield measurable quantities when processed mathematically.

Majorana particle – A particle that is in theory its own antiparticle. Proposed in the 1930s by Ettore Majorana and first discovered in 2012 in Delft.

Many worlds – An interpretation in which the various results of a quantum measurement each exist in parallel realities, meaning that reality is constantly splitting.

Mass – A measure of the amount of matter in an object such as a particle.

Niels Bohr – Danish physicist (1885-1962) and Nobel Prize winner, founding father of the quantum theory of the atom. For a long time, he led the pioneering institute for theoretical physics in Copenhagen, where much of quantum theory was hammered out.

Particle – A component of the building blocks of matter, such as electrons or photons, seen classically as an object with a definite location. In quantum theories it is primarily a measurable manifestation of much vaguer quantum variables, sometimes even wavelike in nature.

Pauli's exclusion principle – A rule in the quantum world saying that no two particles can ever be in exactly identical states. This explains why different chemical elements exist.

Photoelectric effect – The emission of electrons by a metal when light shines on it.

Photon – Packet of light, a quantum of light energy.

Photosynthesis – The storage of the energy of sunlight as sugars and other building blocks by green plants and some bacteria. Quantum vagueness increases the effectiveness of the process.

Planck's constant – A physical constant that determines the size of quantum variables. Named after Max Planck.

Proton – A particle with a positive electrical charge that is the exact opposite of the charge on the electron. The particle itself is much heavier, though. They can be found in atomic nuclei.

Quantum biology – The field of biology in which phenomena such as photosynthesis, enzymes and geomagnetic navigation are related to quantum effects. It is striking that such effects exist in warm and chaotic biological environments.

Quantum calculation (quantum computing) – Using the multiple states of quantum particles in computers that are thereby able to perform huge numbers of calculations in a single step, instead of each calculation taking vast numbers of steps.

Quantum cryptography – Secret communication in which the keys are transferred using the principles of quantum mechanics with no possibility of them being intercepted.

Quantum mechanics – A theory put together by Niels Bohr, Hendrik Kramers, Max Born, and others that formulates quantum phenomena as purely mathematical relationships between measurable variables.

Quantum numbers – Integer values that fully describe the state of a particle or system. They were originally developed to allow atomic spectra to be explained.

Quantum vagueness – The fact that particles or systems are in different quantum states at the same time, until the act of observation forces one of them to be selected.

Qubit – A quantum object that exists in two states at once, unlike a normal bit that can only have a single state at any one time. They are essential for quantum computing.

Schrödinger's cat – A scenario that imagines a cat in an enclosed box, with a quantum mechanical system that determines whether the animal lives or dies. The consequence is that the cat itself is also in a dual quantum state as far as any external party is concerned.

Solvay Conferences – A series of meetings in the first half of the twentieth century that were organized by the Belgian industrialist Ernest Solvay, at which the cream of the scientific world discussed fundamental questions. In the 1920s they were the setting for vigorous debates between Einstein and Bohr about quantum theory. They were always held at the Hotel Métropole in Brussels.

Spin – A property of electrons and other quantum particles that seem to be rotating around their own axis. Unlike a classical rotation, the spin of a particle when measured always has a fixed size and is always either parallel or antiparallel to the measurement device.

Spintronics – An alternative form of electronics that uses electrons' spins rather than their charges.

Spooky interaction (at a distance) – Einstein's term for the fact that separated entangled quantum particles seem able to communicate with each other faster than the speed of light, which would therefore violate the theory of relativity.

Teleportation – Transferring quantum states over a distance using entangled pairs of quantum particles at the two locations.

Tunneling – The phenomenon by which particles such as electrons can pop up on the other side of a barrier, caused by the fact that they are smeared out across space, in quantum mechanical terms.

Uncertainty – A quantum effect that means certain properties of particles cannot be measured precisely at the same time, even in principle. It follows on from the mathematics of quantum mechanics as discovered by Werner Heisenberg.

Wave function – A mathematical function that gives the value of a quantum mechanical property of a particle. The square of that function at any point gives the probability of a measurement finding the particle there.

Wave mechanics – Erwin Schrödinger's theory from 1925 that described the behavior of particles mathematically in wave equations. It gives the probabilities that measurements will show particles to be at a given position or in a given state.

Wave-particle duality – The fact that quantum particles or quantum systems can exhibit the properties of classical waves and classical particles at the same time. That distinction becomes blurred in quantum mechanics.

SOURCES

Jim Baggott, *The Quantum Story: A history in 40 moments*, Oxford: Oxford University Press, 2011.

Martijn van Calmthout, *Einstein's light: een leven met relativiteit* [Einstein's light: A life with relativity], Amsterdam: Contact, 2005.

Martijn van Calmthout, *Einstein in een notendop* [Einstein in a nutshell], Amsterdam: Bert Bakker, 2010.

Martijn van Calmthout, Jelle Reumer, *Nobel op de kaart. Op zoek naar de Nederlandse Nobelprijswinnaars van vroeger en nu (2014)* [Nobel on the map: searching for Dutch Nobel laureates past and present], Hilversum: Lias, 2014.

Brian Cox, Jeff Forshaw, *The Quantum Universe: Everything that can happen does happen*, London: Allen Lane, 2011.

David Deutsch, *The Fabric of Reality*, London: Penguin, 1997.

Richard Feynman, *Six Not So Easy Pieces: Einstein's relativity, symmetry, and space-time*, New York: Basic Books, 1963.

Richard Feynman, *The Meaning of It All: Thoughts of a citizen-scientist*, New York: Basic Books, 1998.

John Gribbin, *In Search of Schrödinger's Cat: Quantum physics and reality*, London: Transworld Publishers, 1984/2012.

John Gribbin, *Schrödinger's Kittens: And the search for reality*, London: Weidenfeld & Nicolson, 1995.

David Kaiser, *How the Hippies Saved Physics: Science, counterculture and the quantum revival*, New York: Norton, 2011.

Jim Al-Khalili, *Quantum: A guide to the perplexed*, London: Weidenfeld & Nicolson, 2003.

Jim Al-Khalili, Johnjoe McFadden, *Life on the Edge: The coming of age of quantum biology*, London: Transworld Publishers, 2014.

Manjit Kumar, *Kwantum: Einstein, Bohr en het grote debat over de natuurkunde* [Quantum: Einstein, Bohr and the great debate about physics], Amsterdam: Ambo, 2009.

Herman de Lang, Vincent Icke and others, *Canon van de natuurkunde: de grootste ontdekkingen en theorieën van 100 belangrijke natuurkundigen* [The canon of physics: the greatest discoveries and theories of 100 important physicists], Diemen: Natuur Wetenschap & Techniek, 2009.

Seth Lloyd, *Programming the Universe: A quantum computer scientist takes on the cosmos*, New York: Vintage Books, 2005.

Chad Orzel, *Hoe leer je natuurkunde aan je hond* [the Dutch translation of 'How to teach physics to your dog'], Hilversum: Lias, 2012.

Abraham Pais, *Subtle is the Lord... The science and the life of Albert Einstein*, Oxford Clarendon Press, 1982.

Abraham Pais, *Niels Bohr's Times, in physics, philosophy, and polity*, Oxford Clarendon Press, 1991.

Alastair Rae, *Quantum Physics: A beginner's guide*, London: Oneworld Publications, 2005.

Ian Sample, *Massive: The hunt for the God particle*, London: Virgin Books, 2010.

Erwin Schrödinger, *What is Life?* Cambridge: Cambridge University Press, 1945.

Leonard Susskind, George Hrabovsky, *The Theoretical Minimum: What you need to know to start doing physics*, New York: Basic Books, 2014.